The Burning House

Unlocking the Mysteries of the Brain

JAY INGRAM

Penguin Books

PENGUIN BOOKS
Published by the Penguin Group
Penguin Books Canada Ltd, 10 Alcorn Avenue, Toronto, Ontario, Canada M4V 3B2
Penguin Books Ltd, 27 Wrights Lane, London W8 5TZ, England
Penguin Books USA Inc., 375 Hudson Street, New York, New York 10014, U.S.A.
Penguin Books Australia Ltd, Ringwood, Victoria, Australia
Penguin Books (NZ) Ltd, 182-190 Wairau Road, Auckland 10, New Zealand

Penguin Books Ltd, Registered Offices: Harmondsworth, Middlesex, England

First published in Viking by Penguin Books Canada Limited, 1994

Published in Penguin Books, 1995

10 9 8 7 6 5 4 3 2 1

Manufactured in Canada

Canadian Cataloguing in Publication Data

Ingram, Jay
The burning house

ISBN 0-14-017878-3

1. Brain I. Title

QP376.I54 1995 612.8'2 C94-931893-0

Phrenology chart by The Granger Collection, New York
Photograph of human brain by Wilder Penfield, Liverpool University Press
Mars photos by Dr. M.J. Carlotto

To all those known only by their initials
whose disabilities have revealed so much
about the brain

Acknowledgments

It goes without saying that anyone who tries to write about the brain is very lucky if he can count on expert help, and I had more than my share. Canadian nationalists should rejoice at the quality of brain research in Canada. Most people have heard of the Montreal Neurological Institute, and it continues the quality work it is renowned for, but internationally recognized research is pursued in many different places. I didn't hesitate to ask the scientists at any of these research centres for help whenever I needed it, which was almost every day.

Two people in particular deserve recognition for their patience in the face of my pestering: Dr. Sandra Black at Sunnybrook Hospital in Toronto, and Dr. Morris Moscovitch at Erindale College, University of Toronto. It might sound paradoxical, but they taught me both how simple and how complex observing the brain can be. There were many others who by their quick responses to questions and requests for publications indicated they too might have been willing to play the role of expert-on-the-line, and they have Drs. Black and Moscovitch to thank for not hearing from me repeatedly. These include John Kalaska, Mel Goodale, Paul Muter, Ronald Melzack, Meredyth Daneman, Brenda Milner and Allan Moffitt. Americans Paul Grobstein and Jonathan Winson lent a hand as well, and

Suzanne Corkin at MIT has been an invaluable contact going back to the days of "Cranial Pursuits" on CBC Radio.

Adrianne Noe at the National Museum of Health and Medicine in the United States updated me on the Yakovlev-Haleem Collection of Normal and Pathological Anatomy and Development of the Brain, and Annette Dukszta escorted me around the Canadian Brain Tissue Bank. Josef DeKoninck deserves thanks again for allowing me to spend a (restless) night in his sleep lab at the University of Ottawa.

Two people whose work figures largely in this book, Mohamad Haleem and Justine Sergent, both died before their time in 1994. I will remember them, as I will a man whom I never met, a stroke patient of Sandra Black's who had lost the ability to recognize faces. He talked to me on the phone shortly before his death about what it was like to see a face and not have a clue whose it was. He was one of the army of stroke patients whose willingness to be studied has added immeasurably to the understanding of the brain.

Working on "Cranial Pursuits" in 1992 with Ira Basen, Chris Grosskurth and Ben Schaub was the inspiration for this book, and they deserve credit for helping me identify what is most fascinating about the brain. Besides, I knew they'd scan this list looking for their names. And of course Cynthia Good, Karen Cossar, Rosemary Reid, Meg Masters and Mary Adachi have all influenced this book. In particular, Meg and Mary have had the unenviable responsibility of clarifying that which was murky, and as usual accomplished that feat without complaining. And thanks too to the Penguin Chapter Title Committee with whom I met over dinner. Thanks as well to David Pecaut who both explored the cathedral square in Milan on my behalf and also offered words of advice and encouragement at the right time.

And finally, once again my family bore the brunt of all this: my children think every father goes upstairs to the computer after reading them bedtime stories. My wife, Cynthia, knows better. I'm looking forward to spending more time watching *their* brains at work.

Introduction

Should I save you some time and trouble right now and tell you what is *not* in this book? There is practically nothing in here about the differences between men's and women's brains, no tricks to improve your memory, no words of encouragement that the aging brain is not declining as you might have feared and not one word about the location in the brain of the centre for sexual desire.

Having established that I have refrained from sensationalism and that therefore my motives must be lily-white, I will acknowledge (without the hint of a blush) that lobotomy, Prozac and repressed memories *are* in this book. So are phantom limbs, a story about a woman who thought her left arm actually belonged to someone else and an account of a man who remembers nothing of the past forty years. Brain researchers turn out stories like these every day, and a book about the brain wouldn't be complete—or as interesting—without them.

But as far as I'm concerned, no account of a person who has suffered some kind of disabling brain injury is complete unless it carries the epilogue: "What does it mean for me?" It *is* bizarre and fascinating that an otherwise rational woman could think that her paralyzed left arm is someone else's, attached to her by surgeons in the hospital where she was recovering after a stroke. But to leave the story there is merely to have walked through the midway glancing at the freaks. Stories like these need reflection, and on closer examination—putting yourself in the patient's place—you will see that her reasoning isn't nearly as strange as it seems at the outset, and that your brain would likely have harboured some of the same thoughts or would have pursued a similar line of reasoning. The human brain abhors a vacuum, and where a story is left unfinished or a puzzle unsolved, anyone's brain will do its best to finish it whether or not it makes sense by so doing.

Defining what's missing after a brain injury is one of the commonest ways to investigate the workings of the brain; another is to look directly at the brain with technologies like magnetic resonance imaging or positron emission tomography. These can produce exciting pictures of the brain at work. As I'm writing this, a set of images of the brains of chess players has just been published in the journal *Nature*. As the players looked at a chess board on a computer screen, they were asked to proceed through a series of tasks, starting with the simple—determining the colours of the pieces—and ending with the most difficult—determining whether checkmate was possible on the next move. In making the latter decision, two main parts of the brain lit up: the frontal lobes and the visual areas at the back of the brain. The explanation offered for this image was that the frontal lobes are involved in planning the checkmate move (a role that they are known to play in other situations) and the visual areas in creating a mental image of the move.

But a word of caution. Seeing areas of the brain light up in a PET scanner is not the final word. These images provide about as much detail regarding the events going on in the brain as satellite photos of city lights on the nightside of the planet reveal about the events on the streets. There is much uncertainty here, as there also is in assuming that there is an exact parallel between the altered behaviour of a brain-damaged patient and the part of the brain that was injured. It may not always follow that the incapacitated part controlled the behaviour that is now missing.

It is both true that researchers have revolutionized the understanding of the brain in the last two decades and that they are still very far from a complete understanding of it. In that sense, brain researchers are a little like Martin Frobisher sailing into Frobisher Bay in 1576 and supposing he was entering "...the West Sea whereby to pass to Cathay and to West India." His was a marvellous accomplishment and the landscape of mountains and ice before him bizarre, foreign and beautiful. But it was not the Northwest Passage. The landscape of the brain that is visible today is in every way just as fabulous as that which Frobisher saw, but neither does it represent the end of the journey. It is not even clear what the end of the journey might look like.

But that uncertainty just makes the whole venture more exciting. The insights into the workings of the brain which are now emerging will not just flesh out or colour what is already known, but will transform it. Take memories as an example. Think of them not as little scenes replayed in your mind, but as moments when many separate parts of the brain which were stimulated during some original event re-create that activity, co-ordinated by an electrical beat, with only the simultaneity of those re-creations giving you the impression of a single remembered scene. That is typical of what's happening to thinking about the brain as research advances.

The brain has been called the most complicated object in the universe, but I think my carpenter/handyman friend James said it better. As we both stood in my kitchen grappling with the very idea of the brain, he broke the silence with: "...Quite the organ."

Contents

part one: GREY MATTER

A Trip to the Museum 1

Reading the Brain 16

Crossing the Synapse 31

part two: NEGLECT

The Cathedral Square 42

Inner Maps of Outer Space 51

The Burning House 60

part three: THE HOMUNCULUS

A Cup of Coffee 72

The Little Man 82

Phantom Limbs 94

"I'm Quite Sure It's Not My Arm" 104

Out of the Body 115

part four: FACES

Do You Recognize This Face? 127

Faces, Cows and Upside-down Dogs 139

The One-Sided Face 148

part five: MEMORY

H.M. 158

Here and Now 176

Then and There 191

Feats of Memory 206

part six: DREAMS

A Freudian Sleep 220

Dreams of Echidnas 232

Additional Reading 244

Index 252

1

A Trip to the Museum

One of the most memorable interviews I ever did was with Mohamad Haleem, the curator of the Yakovlev Collection of Normal and Pathological Anatomy and Development of the Brain in the Armed Forces Institute of Pathology in Washington DC. The setting could not have been better: the Yakovlev Collection is deep inside the only building in that city built to withstand an A-bomb blast. After the Soviets exploded their first A-bomb in 1949, orders went out to build an A-bomb-resistant building in Washington to be the seat of government if there were an attack. Then in August 1953 the Soviets exploded their first H-bomb, instantly rendering the A-bomb obsolete and making the protection useless.

The brain collection itself is in a brightly lit room with an area on one side for the detailed examination of specific brains, and a row of cabinets on the other. The cabinets are the collection's version of the heavy library shelves on rollers that move when you spin a hand crank. But there are two differences here: instead of book shelves, each of these cabinets is stacked with sixty or seventy drawers a couple of centimetres deep.

The second difference is that the Yakovlev Brain Collection has a specially designed custom-made electrical system to propel the cabinets—no hand cranks here.

The second thing I had going for me was the interviewee, Mohamad Haleem. He was curator of the collection, but more important, he had worked side by side for more than twenty years with the founder, Dr. Paul Ivan Yakovlev. Yakovlev died in 1983, but Mr. Haleem still spoke of him with deep reverence (called him "the master") and treated the collection as if it were a sacred trust, which in a sense it is. This brain museum is a serious scientific project, and is visited by neuro-experts from around the world. It's not just that there is an incredible variety of brains available here for study, it's also that each brain has been sliced in exactly the same way, in a "gapless series," the whole loaf sliced from crust to crust. The number of slices from a single brain can easily top four thousand, although the exact total depends on whether the brain was cut from front to back, top to bottom or side to side. The important thing for visiting scientists is that you know what you're getting when you go to the Yakovlev Collection. If you want to measure the dimensions of a part of the brain like the cingulate gyrus in the brain of a stroke victim, and then compare that brain, slice for slice, with another, you can. You can examine slices from a ten-week-old fetal brain or from a hundred-and-one-year-old one. In the collection are brains damaged by carbon monoxide poisoning, late-stage syphilis and Alzheimer's disease, and brains from monkeys, a sperm whale and a horse.

Although brains are offered to the collection all the time, only the most unusual are accepted. (Unusual in this instance might even mean the closest thing possible to a "normal" brain.) It takes a year for a donated brain to be chemically hardened to the point where it will slice perfectly without being squeezed out of shape by the blade. Each brain is cut into slices about one-thirtieth of a millimetre thick (four or five of them stuck together would equal the thickness of a single page of this book); each slice is protected between two thin sheets of glass the size of a small picture frame. The slices are thin to the point of transparency, and so are easily viewed in the microscope.

Far from being sensational, the Yakovlev Collection is a museum in the traditional sense: a collection painstakingly gathered and standardized to be of use to investigators now and in the future. But of course there is a sensation associated with being surrounded by hundreds of thousands of slices of brain. For most people it would be queasiness, although just why that is I'm not sure. Maybe it's the texture—brains are wet, soft and heavy, and the idea of actually holding one repels most people. Of course in the Yakovlev Collection this is not a major problem, given that the brain slices in the drawers are about as far removed from a "wet" brain as processed cheese slices are from the original product. Anyway, brains aren't special in that sense—my wife, Cynthia, is convinced that holding a kidney would be worse.

The slight sense of eeriness that brains induce probably stems from the fact that for most of us the brain is more than just a body part—we associate a person's brain directly with their thoughts and feelings, and so when holding the brain we have the person "in our hands." That provokes the kind of emotional experience that we likely wouldn't enjoy had Dr. Yakovlev collected three-metre sections of people's intestines and called it the Yakovlev Duodenum Collection. Having no particular reverence for the brain nearly proved the undoing of one group, the Fore people on the island of New Guinea in the South Pacific. They routinely carried out rituals during which any newly deceased relative was consumed. Unfortunately the tribe was carrying a rare infectious agent that attacked the brain, causing a slow-developing but usually fatal movement disorder. Anyone who ate part of the brain of someone who died from this disease, called "kuru," more or less ensured that they too would succumb. As soon as the tribe was encouraged to discontinue this particular part of their rituals, kuru faded away, although before it did, the brains of two kuru victims made it into the Yakovlev Collection.

Although today the brain has achieved a special prominence, this has not always been the case, even in societies that devoted a great deal of time and energy to thinking about it. The ancient Egyptians, although they had elaborate concepts relating to mental and spiritual life like the *ba* (a non-material representation of the character or personality) and the *ka* (a ghostly sort of double which outlived its mortal counterpart), believed that these elements resided in the heart, not the brain. During mummification, the heart was carefully preserved; the brain was sucked out through the nostrils or the base of the skull and thrown away.

From the Roman Empire until a few centuries ago, even though anatomists had by then realized that the brain was the organ of thought, they misguidedly assigned the key role to the fluid-filled spaces in the brain called ventricles, and dismissed the brain cells and their connections as mere packing material. Even specific mental attributes like reason, judgement and memory were assigned to different, though connecting, spaces. There is no known fluid that could embody the complexity of the brain and today the cells of the brain and their connecting fibres—that which was dismissed as packing material—are considered to be the essence of the person. Even so, any discomfort at the thought of being surrounded by brains is irrational no matter how you look at it. If you're spiritually inclined, you'll believe that the soul has long since departed the brain before it was added to the Yakovlev Collection. If on the other hand you think death brings everything to a halt, why should handling a brain make you any more squeamish than holding a thighbone? It's all just dead tissue. Whether it makes sense or not, brains are something special, and doing a radio interview surrounded by sixteen hundred of them promised something memorable. For good measure Mr. Haleem had selected for my benefit some of the most interesting specimens to display on the light cabinets.

Among them was one from a lobotomy patient labelled "Male, fifty-nine years old. Post-operative survival, six months."

Because lobotomy was an operation of the 1940s and 1950s, most of us have only the vaguest idea of what it is, how commonly it was performed, and certainly have no clue as to the high regard in which it was held. It called for slicing clean through the frontal lobes of the brain and was prescribed for psychiatric patients who were deemed to have no other hope of reducing their anxiety or anti-social behaviour; thousands of these operations were done. Evaluations of lobotomy patients published in the 1940s were unanimous that the majority had benefited from the operation. At least according to medical experts, these patients were no longer disabled by the tortured thoughts or unacceptable behaviour of schizophrenia or manic depression. As bizarre as it sounds today, the Portuguese doctor who was responsible for introducing lobotomy to the world, Egas Moniz, shared the 1949 Nobel Prize for Physiology and Medicine for "his discovery of the therapeutic value of prefrontal leukotomy (lobotomy) in certain psychoses." There are estimates that 40,000 of them were done in the United States alone, mostly in the early 1950s. One of the reasons that this crude form of treatment for psychiatric disorders was embraced overenthusiastically—and certainly before adequate evaluation—was that there were countless World War II veterans with severe psychiatric problems, and treatment was difficult: too few psychiatrists and very little in the way of antipsychotic drugs.

Many people have the misconception that a lobotomy involved actually removing the front of the brain and throwing it away. Although there was one technique in which pieces of the frontal lobes were removed (and weighed to record the exact amount taken out), most of the time they were just *disconnected*, usually by inserting a blade into the front part of the brain and then swinging it back and forth pendulum-style. This lobotomized brain in the Yakovlev Collection had been sliced (after death) to provide a side view which profiled the incision that had isolated the frontal lobes but left them in their place, held there by the bones of the skull.

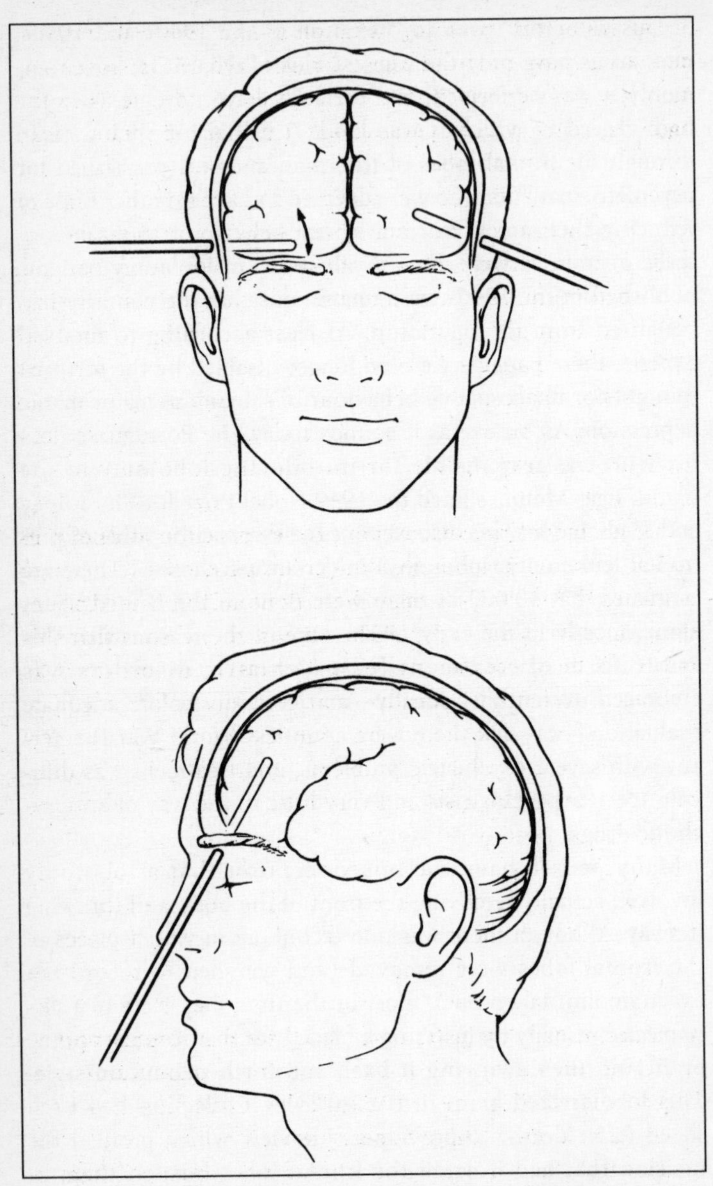

An illustration of one lobotomy technique

The history of lobotomy helps illuminate the terse notation accompanying some of these brains. One is listed in the Yakovlev catalogue as "C-64...32 ys...13 ys...F...H." Female, aged thirty-two, post-operative survival thirteen years, sliced horizontally. And this specimen? A lobotomy in 1951—at the very height of the lobotomy movement—for anorexia nervosa. Mr. Haleem also selected for display a brain from a stroke victim, another from a four-year-old with Down's syndrome and several from animals, including a gorilla and a dolphin.

Maybe I just got too used to the brains in the few minutes I had been there; maybe I thoughtlessly assumed that working in here on a daily basis would ensure that you'd develop a casual approach to the remains. Whatever it was, I overstepped my interviewer's bounds a little when Mr. Haleem and I began to discuss his mentor Paul Ivan Yakovlev. When Yakovlev died in 1983, his brain immediately joined the collection as one of its star specimens. It is, of course, sliced like all the rest...well, not exactly. Mr. Haleem bent the rules for his boss's brain by staining (for viewing through the microscope) every tenth slice instead of the usual one in twenty. The result is more slices for Dr. Yakovlev than for any of the other brains, and more drawers for those slices. (Like an art gallery, only part of the collection of slices is on display, with the rest kept in storage.) Mr. Haleem also took care to slice Dr. Yakovlev's brain in the direction "the master" had preferred: in sagittal section, that is, sliced from left to right. I was fascinated—I wanted to see this brain:

Ingram: His brain is here?
Haleem: Absolutely.
Ingram: Can we see it?
Haleem: Absolutely. It's here. You'll see it of course, and I'll be very sad of course, but I'm going to show you the brain. Actually, this is Dr. Yakovlev's...brain...
Ingram: It doesn't have a special place?
Haleem: No. He is just another patient who unfortunately ended up here.

Ingram: His name isn't even on it.
Haleem: No, we cannot in fact give out names.
Ingram: Just numbers.
Haleem: In fact in the medical profession you cannot talk about names and maybe we should not talk about names.
Ingram: Okay, so he's just MU...
Haleem: MU 157.
Ingram: 1183. Eighty-three being the year of his death?
Haleem: That's right. These are, as you can see, sections from his brain. This is number 511. For me, for sentimental reasons or maybe other reasons, I decided to stain every tenth rather than the standard every twentieth. Because I want as many sections as I can from this brain. And, uh, here it is. You can see here how many trays. Approximately...about...thirty-five trays from that one brain.
Ingram: If you look at this brain as a sample brain in this huge collection, does it look different to you?
Haleem: Absolutely. Actually, as I'm talking to you, I'm really trembling. Honestly. I feel...when I am talking about other brains I'm fine. But this is a special thing for me. Even if you're not here, every time I pass by here I *feel* somehow different.

The average human brain weighs about 1,400 grams, with men's brains outweighing women's brains—on average—by about 100 grams. (But detailed anatomical studies have established that the number of brain cells or neurons in every cubic centimetre of the female human brain is greater than in the male, the net result being that the total number of neurons is about the same in both sexes. Remember that the next time you hear someone trying to explain why men's brains are bigger.) It's a pretty solid 1,400 grams, three pounds or so—a human brain has a good heft to it—and the natural colour is more pink than grey. The really surprising thing about the appearance of a living brain (of which most people are likely to get their first glimpse on The Learning Channel) is the size and prominence of the blood vessels running across the surface; I knew they were the supply lines for the brain's enormous requirements for oxygenated blood, but I had no idea that they bulged from the brain's surface like the veins on a body-builder's arms. It isn't really that surprising when you consider the requirements: about 25 per cent of the body's entire oxygen intake is shunted off to the brain, where it fuels the electrical activity of brain cells.

The other important feature of the brain's surface is one that you've undoubtedly seen, even if you haven't given it much thought: the folding of the surface, especially on the sides and the top. The whole outer cap of the brain, the cerebrum, is an endless series of low rounded hills separated by deep narrow crevasses. Each hill is a "gyrus"; each valley a "sulcus." Some of the main ones retain the name of the anatomist who first called attention to them: the Fissure of Rolando (eighteenth-century Italian), the Sylvian Fissure (seventeenth-century French), but most simply bear labels of location, like the "prefrontal" gyrus. This same geographic approach is used to divide the entire cerebral surface into four lobes: frontal, parietal, temporal and occipital. These areas were chosen to correlate to the divisions in the skull directly above, but the match is only approximate. The frontal lobes are the largest, occupying about 30 per cent of the entire cortex: they extend forward from just in front of the ears. The parietal lobe comprises the crown of the brain, extending back about to where the sweatband of a baseball cap would lie. The occipital lobe is at the back of the skull, and the temporal lobes spread out around the ears.

There are three closely related terms that I might as well sort out here and now. The cerebrum is the convoluted outer surface of the brain that is divided into the four lobes mentioned above. The dividing lines between the lobes are subtle and somewhat arbitrary, but there is a clear-cut division in the cerebrum: right down the middle. The human brain as seen from

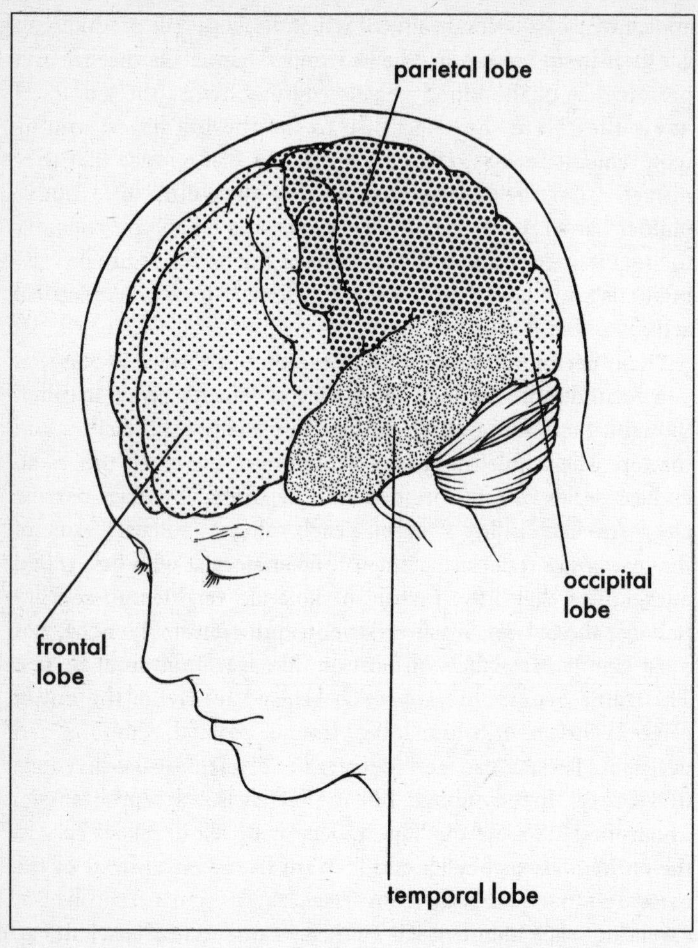

above looks like a walnut, with the left and right halves called the cerebral hemispheres. And finally, the surface layer of the whole cerebrum, both right and left hemispheres, is called the cerebral cortex, cortex meaning "bark." It is only a few millimetres thick, but all of the brain cells in the entire cerebrum are found in that thin superficial layer. The space below in the interior of the brain is taken up by the so-called white matter—cabling connecting one part of the surface with another.

Crumpling up the cerebral cortex makes it possible to fit much more of it onto the brain. If you smoothed the cortex down flat, it would hang over the edges of the brain. The only way it will fit inside the skull is to scrunch it together. That has been one of the main features of the brain's evolution over the last million years: the addition of more deeply folded cerebral cortex, especially in the front.

I once read that if the folds in the cerebral cortex were smoothed out it would cover a card-table. That seemed quite unbelievable but it did make me wonder just how big the cortex would be if you ironed it out. I thought it might just about cover a family-sized pizza: not bad, but no card-table. I was astonished to realize that nobody seems to know the answer. A quick search yielded the following estimates for the smoothed-out dimensions of the cerebral cortex of the human brain.

An article in *Bioscience* in November 1987 by Julie Ann Miller claimed the cortex was a "quarter-metre square." That is napkin-sized, about ten inches by ten inches. *Scientific American* magazine in September 1992 upped the ante considerably with an estimate of 1 1/2 square metres; that's a square of brain forty inches on each side, getting close to the card-table estimate. A psychologist at the University of Toronto figured it would cover the floor of his living room (I haven't seen his living room), but the prize-winning estimate so far is from the British magazine *New Scientist*'s poster of the brain published originally in 1993 which claimed that the cerebral cortex, if flattened out, would cover a tennis court. How can there be such disagreement? How can so many experts not know how big the cortex is? I don't know, but I'm now on the hunt for an expert who will say the cortex, when fully spread out, will cover a football field. A Canadian football field.

Cases like this make me reluctant to relate other fascinating claims about the brain for fear they might prove equally unreliable. One such story is that if you allow a fresh brain to sit quietly—in a pasta bowl would be perfect—it will slowly "slump" and gradually lose its upright shape. It will flow because it contains so much liquid. I have no idea where I originally read this, and I'm sure there are very few people who have actually watched what happens to a brain that's left alone in a pasta dish. On the other hand, the cerebrum is about 85 per cent water, and before a brain can be sliced for the Yakovlev Collection, it has to be infused with plastic and hardened.

If the cerebrum is just rumpled up like a rug caught under the leg of the sofa it's hard to imagine that there would be any significance attached to being in the depths of a crevasse rather than on the crest of a hill, or even on one crest instead of another, but there have been a couple of recent findings suggesting that there might indeed be some unsuspected significance to the brain's hills and valleys.

In one series of studies Dr. George Ojemann and his colleagues at the University of Washington in Seattle have tried to map with great precision the language areas on the surface of the left hemisphere, and they have found that there is tremendous variability from one patient to the next in the exact location of areas important for naming things. One of his most intriguing and least understood discoveries has been that one of these differences is correlated with skill in language. Key naming areas are found on top of one gyrus in those who were good at language, and on the next one over in those who weren't.

On the other hand, a study of memory published early in 1994 by scientists in Toronto revealed that some of the most intense brain activity was in the depths of the sulci, the deep crevasses between folds, rather than on top of the folds themselves. Volunteers in these experiments listened to a series of sentences, many of which they had heard the day before. Brain-scanning images of the activity in their brains when hearing day-old sentences (and presumably remembering them) showed that most of the activity was in the front of the brain, centred on the crevasses. Like George Ojemann, these investigators don't understand the significance of that discovery, but such studies might be opening the door to thinking about mapping the topography of the brain as well as its latitude and longitude.

This quick look at the outer surface of the brain reveals little of why it is a marvel of nature. But you wouldn't have to dig very deep to begin to glimpse what makes it special. A microscope reveals that the convoluted cerebral cortex is packed with brain cells, both in layers (at least six in some areas and maybe everywhere), and columns, side by side. Below that lie the fibres that make the connections among brain cells, whether neighbouring or distant. These subsurface circuits move information from one part of the brain to others, and below them lie other nerve centres, more "ancient" in the evolutionary sense than the cortex. Here deep in the brain is where you find the so-called mammalian and reptilian brains which house centres of emotion, drives such as thirst and sex, and regulatory networks for breathing and sleeping. These are the "old" parts of the brain that we share with other animals, but the real human brain work, the computations and processing of information that we are apparently able to do like no other species, belongs to the thin surface layer of the cortex. It sets us apart from all other living things, even after death.

The precision slicing and staining of the specimens in the Yakovlev Collection is only one way to store brains. In Toronto, the Canadian Brain Tissue Bank collects brains at least as rapidly as the Yakovlev Collection, but like all banks turns deposits into withdrawals at top speed. Researchers interested in chemistry rather than anatomy submit a wish list to the CBTB and when an appropriate brain arrives, it is divvied up into pieces—twenty grams of temporal cortex here, a couple of grams of substantia nigra there—and transferred to the researchers as quickly as possible. The fresher the brain the better, because research into brain chemistry requires tissue that is as close to the living state as possible. The bank does have the equivalent of safety deposit boxes: half of each brain is preserved in formaldehyde, and the floor of a room in the Banting Building on College Street in Toronto is littered with Tupperware containers containing chunks of brain.

Part of the incentive for collecting brains in the first place was the hope that some structural feature could be identified—especially in famous brains—that would account for outstanding abilities. But it has never happened. While I was writing this book, the director of the Moscow Brain Institute, Oleg Adrianov, reported that seventy years of study of Lenin's brain had culminated in the finding that it contained "nothing sensational."

The Russian investigators who studied his brain must have had the faith that something unusual would turn up, some brain feature that would shout, "This is what made Vladimir Ilyich great!" But differences in brain structure have never been successfully correlated to achievements. For instance, some scientists today are busy trying to prove that the size of the brain correlates with intelligence. In one of the most reliable of such studies, Nancy Andreasen of the University of Iowa and her colleagues compared intelligence (as measured by IQ scores) to the size of different parts of the brain. They found that bigger brains are more likely to produce higher IQ scores, but they emphasize that the correlation is "modest" and that "many other factors must also be important." In other words, you can't look at one brain in isolation and calculate the intelligence of its owner from the size. The same goes for crevasses, folds and all other structural features, which casts a long shadow on Dr. Adrianov's claim that Lenin's brain, although exhibiting nothing dramatic, was neverthless "undoubtedly the brain of a talented man."

Einstein's brain still soaks in formaldehyde in the Lawrence, Kansas apartment of the man who performed his autopsy in 1955, Dr. Thomas Harvey. There has never been anything of significance found in the brain, and at 1,230 grams, it wasn't particularly big. Apparently there is not even any great demand to transfer it from Dr. Harvey's apartment.

Ingram: Mr. Haleem, are you going to have your brain added to the collection?

Haleem: This has been my desire and of course the

> influence of Dr. Yakovlev and what I have been really doing for the past thirty years, serving the scientific community. I think the least I can do is, well, add one more contribution.

Those words, spoken in 1992, now have a poignancy that wasn't apparent at the time. Mr. Haleem died from cancer early in 1994, and he likely knew he had the disease when I interviewed him. Happily, his wish to have his brain added to the collection has been fulfilled, and his brain is now in the long process of preparation for sectioning. Unlike his mentor, Dr. Yakovlev, Mr. Haleem apparently had no particular preference for the orientation of slicing, but he indicated the number he'd like assigned to his brain slices, and the exact shelf on which they are to be stored. Dr. Adrianne Noe, who at the time of writing was in charge of the collection, is one of three people who conferred with Mr. Haleem about the preparation of his brain, and she told me that in all her career of caring for large museum collections, she had never before participated in a discussion about the preparation of a specimen with the specimen itself. One final note: the collection is now known as the Yakovlev-Haleem Collection of Normal and Pathological Anatomy and Development of the Brain.

2
Reading the Brain

A brain museum is a nice place for browsing and is an invaluable archive, but you can't tell very much about how a brain works by examining it whole or even sliced. If the brain is really what makes us who we are, it would be preferable to see it in action, directly or indirectly.

In the early nineteenth century a theory of the brain (which soon escalated into a fad of grand proportions) argued that you could infer what was going on inside simply by examining the surface—not of the brain, but of the *skull*. This discipline, called phrenology, was based essentially on a two-stage argument: first, that the brain is organized into autonomous regions which bear the responsibility for certain personality characteristics, and second, that those brain regions, if particularly well developed in any individual, would actually push up on the skull from below and create a bump that you could feel on the top of the head.

Wonder, Wit, Tune, Language and *Vanity:* if a person were outstanding in any of these, or in any of the thirty-odd others identified by the phrenologists, then he or she would supposedly

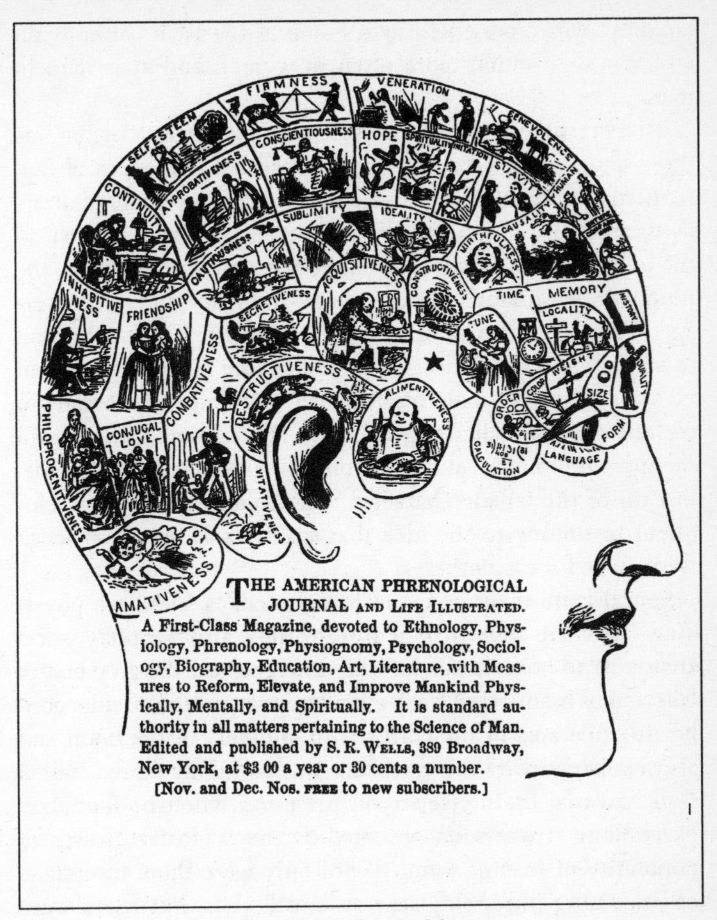

The phrenological brain

have a palpable lump in the skull directly over the part of the brain thought to be responsible for that attribute. *Individuality*, for example, was positioned just above the nose and was thought to be the part of the brain responsible for forming ideas. (Apparently this claim was based on someone's observation that Michelangelo's skull had a bump in this area but the Scots—as a group—didn't.) *Philoprogenitiveness*, or the love of children, was represented by a bump at the back of the brain and was seen much more often in women and apes than in men.

It wasn't simply that the person with the largest "organ" (as these brain areas were called) possessed the maximum of that particular quality; each individual's brain was considered on its own, and the prominence of each organ was rated relative to the others in that brain. So while Johan Kaspar Spurzheim, the number one promoter of phrenology, could agree that, on average, women's brains were smaller than men's, he would go on to argue that it didn't necessarily follow that all qualities would be stronger in the male brain than the female: "the organs of secretiveness and circumspection [in the female brain] are still the most prominent, and thus contribute essentially to the formation of the female character." His comments stand as eloquent testimony to the idea that sexism and pseudo-science were made for each other.

Spurzheim's mentor, Franz Joseph Gall, a Viennese physician, was the inventor of phrenology and a pretty good anatomist to boot: it was he who differentiated the grey matter (the actual brain cells) from the white matter (the circuits connecting one area of the cortex with another) in the brain and also explained why the cerebrum was so highly folded. But if Gall had real brain science on his mind when he launched phrenology, it was soon obscured by the explosive growth in popularity of reading bumps—not only were there thousands of practising phrenologists you could visit, but there were dozens of societies and even several journals devoted to its study.* Societies and journals notwithstanding, in the long run phrenology couldn't survive the easily demonstrated lack of correlation between intelligence and personality and the shape

* Not that having a journal is necessarily meaningful—after all there is a scholarly journal in the United States devoted to the study of Klingon, the language spoken by the aliens of the same name in the television series "Star Trek." It is called HolQeD ("linguistics" in English) and carries articles with titles like, "First Steps Towards a Phonological Theory of Klingon."

of people's heads. But it was a slow death: while popular interest in phrenology was already beginning to flag by the middle of the nineteenth century, surveys showed that even as late as 1925, 40 per cent of Americans still believed in phrenology; that figure finally dropped to 0 per cent by 1974. The British Phrenological Society wasn't disbanded until 1967.

Phrenology had always been under attack by scientists, was being openly ridiculed in print and cartoons in the late eighteen-hundreds and today is seen as an absurd sort of neurological mass delusion, but we shouldn't be so quick to ridicule the theory of phrenology as nonsense that could only have come from the ignorance of the past. Today neuroscientists are far from being phrenologists, but they are perfectly willing to localize a variety of mental functions to certain parts of the brain: the recognition of faces happens largely in the right hemisphere; proper names are thought to be located in, or at least routed through, the tip of the left temporal lobe; and working or short-term memory operates mostly at the front of the brain. The idea that different kinds of mental operations happen in different places is as important now as it was for the phrenologists. Of course, no one today dares to give anything as specific as philoprogenitiveness a place in the brain, nor would anyone claim that somehow these areas of the brain, if well developed, will push up the skull immediately above, but the principle is not dissimilar.

Even the idea of looking at bumps isn't so ridiculous. When scientists studying the evolution of our species tackle difficult questions like "When did our ancestors first begin to talk?" or even "When did the first humanlike brain appear?" they have to dig deep to come up with any kind of evidence—literally. It is possible to re-create the brain of one of our extinct ancestors by looking at bumps on the surface of a fossil skull. But it's not phrenology—in this case it's the *inside* surface of the skull, and it is really not so much bumps as faint grooves or depressions in the bone that capture the interest of anthropologists. They, of course, are looking at skulls belonging to those of our ancestors who are considered good candidates for the creators of human language. Skulls more than two million years old are just too small to have housed brains capable of the complexities of language, while anything more recent than two hundred thousand years or so is presumed already to have language well in hand. It's the 1.8 million years in between that's of interest, because somewhere in that period the capacity for language appeared, took over parts of the brain and grew, both in importance and in size.

The brain presses against the inside of the skull so firmly that some of the hill-and-valley organization of the gyri and the sulci and also the most prominent blood vessels on the brain surface leave shallow impressions in the bone. As the human brain grew dramatically over a period of a million and a half years, some parts of the brain expanded more rapidly than others, and the changing patterns left by those areas can be mapped by looking at a chronological series of skulls. An expert can search for signs of the appearance or early development of the frontal lobes, which are greatly expanded in our brains, or even the left-hemisphere language areas that are prominent today in the modern human brain.

If a fossil skull is found in several pieces, as most are, it can be glued together temporarily to allow a silicone rubber mould to be applied to the inner surface of the skull. Once that is laid down, the entire skull is filled with plaster of Paris. When the glue holding the fragments of skull together is removed, and the skull disassembled, what remains is a rubber-coated plaster of Paris hominid brain whose surface detail reflects that of the original. Then the scientists gather round and speculate, and while it isn't phrenology, it must *look* like it—brain experts with electronic sliding calipers measuring the convexity of the parietal lobes and the breadth of the frontal lobes and making pronouncements about the mental capacity of the brain's original owner. More often than not, the experts disagree about what they think they see. The impressions left on the inside of the skull are just shallow and indistinct enough to allow plenty of room for interpretation. And where there is room, academics will fill it. One of the most heated arguments in this area raged over the antiquity of the human type of brain organization. At issue was a faint line on one of the most famous fossil skulls of all time, the so-called Taung child, found in South Africa by Raymond Dart in the mid-1920s. Dart had realized at once that this was an entirely new animal, and gave the infant (complete with baby teeth) the formal scientific name *Australopithecus africanus*, but called the child colloquially a "man-ape." He was convinced of a close relationship between Taung and modern people because of the infant's humanlike brain organization, visible as lines on the inside of the skull. In 1980 palaeoanthropologist Dean Falk suggested that Dart had made a major-league error—he had mistaken a line marking the joint between two bones of the skull for the mark left by one of the major sulci or grooves in the cerebral cortex. According to Falk, this misidentification led Dart to assert, incorrectly, that the Taung child's brain was more human than ape. For the next few years Falk had to defend her unexpected claim against attacks from her academic rivals, but it seems now as if she has gained some ground as many now buy the idea that the Taung child has a brain like an ape. But this debate isn't over even yet, and unfortunately it is almost always true that the brain imprints on fossil skulls are ambiguous and inconclusive.

Even so, it is possible to speculate. In Kenya in the mid-1980s the nearly complete skeleton of a twelve-year-old boy belonging to the species *Homo erectus*, a species considerably closer to us than any Australopithecine, was discovered. The boy had lived his short life about one million six hundred thousand years ago. The skull was largely preserved, and the brain has been reconstructed from it. There are subtle asymmetries between the right and left hemispheres which suggest that the boy was probably right-handed, and a well-developed area on the front left-hand side of the brain that might have played a role in language and/or the co-ordination of complex hand movements. Reading bumps to illuminate a fossil brain like this one is still something to be pursued with great caution, but even at that it makes more sense than phrenology.

It's too bad in one sense that phrenology didn't work out, because it would have been perfectly non-invasive and even entertaining way of studying the brain. But it still would have had some shortcomings. Even if you could argue that certain areas of a brain responsible for particular qualities are better developed than others, that doesn't tell you very much about how that brain actually works in day-to-day situations. What happens when a brain noted for its philoprogenetiveness has to balance the cheque-book? Does a brain with a prominent organ for republicanism react differently in the voting booth than others? These are questions phrenology could not have answered. By the late 1920s the search for some technique that could reveal the living brain at work had turned to electricity and an instrument called the electroencephalograph, the EEG.

It might sound strange to be searching for electricity in the brain, especially in the hopes of identifying it with thought processes, but there had been suspicions of a link between nerves and electricity since the 1770s when an Italian scientist named Luigi Galvani showed that a disembodied frog's leg would kick if it were hung from a brass hook with the foot touching a silver plate. Galvani thought he had discovered that animal tissues generate electricity, and while it turned out he was right, this experiment does not demonstrate that phenomenon. In this case, the dissimilar metals generated a weak electric current that caused the main motor nerve to fire, commanding the leg muscles to contract. Nonetheless Galvani had opened the door to the study of the electrical nature of life.

About a century later electrical activity was first recorded from the exposed surface of the brains of dogs and rabbits, but it wasn't until the mid-1920s that the idea of measuring the electrical activity of the brain was extended to humans. A German psychiatrist named Hans Berger succeeded in recording his son's brain waves using only metal plates strapped to the head as detectors. Although the boy was probably more delighted than anyone that there had been no need to remove part of his skull to detect the electricity in his brain, the success of this non-invasive technique also meant that the electroencephalograph (the "electric head writer") could be used without fear of risks to health.

Even so, the EEG recordings that Berger published—simple series of long wavy lines—made little impact on the scientists who saw them, apparently because most of them assumed that something like this couldn't produce meaningful results; in their minds it was absurd to expect that the complexity of the brain could be reduced to something simple and concrete like an EEG. Attitudes towards the electroencephalogram have changed, but brain scientists are still trying to understand exactly what the wave patterns mean: they vary from choppy and confused to slow and rolling, and while it is supposed that these waves represent the collective electrical activity of millions of brain cells (sensed as they are from the relatively remote position of the skin overlying the skull which in turn shields the brain), explaining why they should be synchronized like this and what it all means hasn't been easy. But the EEG has always been intriguing if not revealing: even in the late forties and early fifties, scientists like the American EEG pioneer William Grey Walter were seeing differences in wave patterns from person to person that he called "brainprints." In the June 1954 issue of *Scientific American* magazine, after pointing out that the EEG machines were by then so complicated that each contained "dozens or even hundreds of radio tubes," Grey Walter went on to discuss what the EEG had discovered about the infant's brain: "at birth the brainprints of infants are generalized, but at an early age, around *three* or *four* [italics are mine], the child's brainprint acquires the individualistic features of an adult's." Grey Walter described how different kinds of rhythms make their first appearance at different ages: theta rhythms (about five or six waves per second) at about a year, alpha rhythms (nine to eleven per second) during the second and third year.

Forty years later, the EEG presents a different picture of the one-year-old's brain. At the Salk Institute in La Jolla, California, just outside San Diego, Helen Neville and her colleagues measure brain waves too, but the big difference between their work and Grey Walter's is precision. In studying the language competence of very young children, Dr. Neville can use computers to time very precisely the sudden bursts of brain electricity recorded by the electrodes on their heads, pinpoint the location of the bursts and link them to the appearance or sound of a single word. One of the most intriguing things that can be done with this technology is to watch what happens inside a child's brain as she listens to single words in a set of headphones. At twenty months of age, these precisely timed versions of the EEG reveal that the brain is in the process of changing its organization to accommodate language. Within a tenth of a second of hearing or seeing a word, the brain reacts electrically, but there are differences depending on the child's language skills. In what might seem to be a paradox, the brains of twenty-month-old children who aren't good with language react to the sound or appearance of words with a widespread high-energy electrical response over both hemispheres, while in the brains of linguistically advanced children those same bursts of electricity are less energetic and much more restricted in area. Bigger isn't necessarily better when it comes to the brain; as this and other studies have suggested, the brain gets more efficient and more tightly focused as it learns.

At about the same time as electrodes attached to the scalp were recording patterns of changing electrical activity in the brain, the Canadian neurologist Wilder Penfield was working the other side of that coin by stimulating the brain electrically and observing a wide range of reactions, ranging from the interruption of language to the apparent reawakening of long-lost childhood memories. Many of Penfield's patients in Montreal were epileptics whose seizures were not adequately controlled by the available drugs, and had decided on brain surgery as a last resort. It might sound Draconian (or Frankensteinian) to remove a piece of brain tissue as a medical procedure, but you must not forget that the patients who were considered for this were having their lives completely disrupted by their epilepsy. If removing a scarred piece of brain tissue eliminated the trigger for seizures, most of them were perfectly willing to tolerate the consequences, and indeed removal of brain tissue for the control of epilepsy is still a relatively common surgical procedure.

Penfield and his surgical colleagues often had good reason to target a particular part of the brain for removal, based on EEG recordings, qualities of the seizures that could be traced to that part or even evidence of a specific wound to the brain, but they still had to minimize the likelihood of damage to important brain areas, especially those controlling language. Losing speech would be too high a price to pay for relief from seizures. If the surgeons were planning to take a piece of brain tissue from the speech hemisphere (almost always the left), they would conduct a series of tests to locate and mark areas responsible for speech, taking care to avoid them with the scalpel. What's most amazing about this testing is the way it was done. The patient would lie quietly on a bed in the operating room as the side of the skull was removed and the membranous surface lining of the brain peeled back. Because the brain itself has no pain receptors, the patient could remain fully awake with half her brain exposed and be completely unfazed. Then she would be shown pictures or the names of a variety of objects and be asked to identify them out loud. At the same time, Penfield would use the tip of an electrode producing a weak electric current to roam over the surface of the patient's brain, touching it lightly on one spot as she was answering a question.

If the patient were suddenly unable to answer, or had trouble spitting the word out, or repeatedly came up with the wrong word only to correct it as soon as the electrode was removed, the medical team concluded that the electrode had been resting on an important speech area. A small square of sterile filter paper would be dropped on the spot, and Penfield and his electrode would move on. Eventually the speech areas would be covered with little squares of paper, and the surgeons could proceed with the surgery confident that they would damage language as little as possible. Of course those little squares of paper also formed a map of the main speech areas in the brain, and allowed Penfield to confirm in a direct way what psychologists had suspected for decades: that there were (almost always) two main speech areas in the left hemisphere.

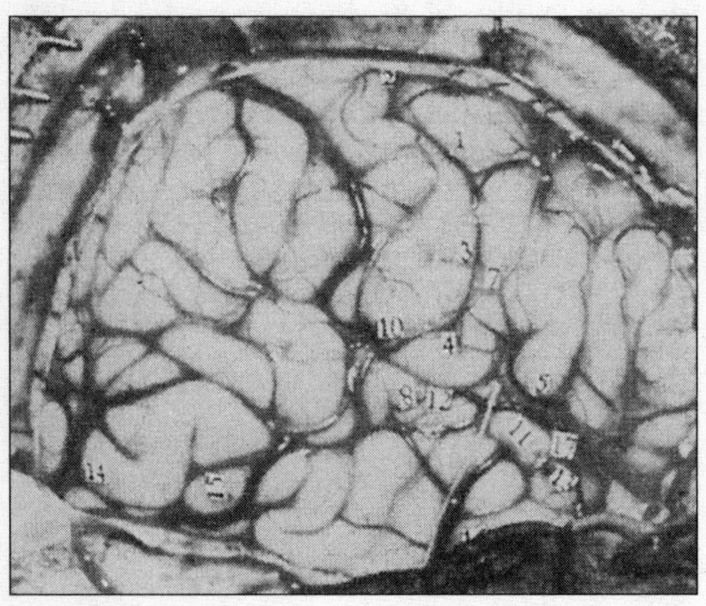

M.M.'s exposed brain as Wilder Penfield saw it.

Outlining the speech areas was fascinating to psychologists and brain surgeons alike, but Penfield's most dramatic discoveries had nothing to do with speech. In patient after patient, this exploration of the surface of the brain with the live electrode apparently succeeded in awakening long-lost memories. This grabbed everyone's attention. One female patient who had come to Penfield for surgery to treat her epilepsy explained that just prior to a seizure, and while retaining awareness of the present, she would feel as though she were back in her past, at a railroad station in a small town which she said could have been Vanceburg, Kentucky, or possibly Garrison. "It is winter and the wind is blowing outside and I am waiting for a train." These preseizure flashbacks were often accompanied by feelings of fear and also *déjà vu* "as though I had lived through all this before." When Penfield explored the surface of her temporal lobe, where it was suspected these seizures began, she had some startling responses to the touch of the electrode:

> Stimulation at Point 15: Just a tiny flash of familiarity and a feeling that I knew everything that was going to happen in the near future...as though I had been through all this before and thought I know exactly what you are going to do next.
> Point 11: I think I heard a mother calling her little boy somewhere. It seemed to be something that happened years ago.
> Point 11 again: Yes I hear the same familiar sounds, it seems to be a woman calling, the same lady. That was not in the neighbourhood. It seemed to be in the lumber yard...I've never been in the lumber yard.
> Point 13: Yes I hear voices. It is late at night, around the carnival somewhere—some sort of travelling circus. Then after the electrode is removed: I just saw lots of big wagons that they use to haul animals in.

This patient, known as M.M., wasn't the only one who had specific childhood memories that could apparently only be brought back by the touch of an electrode or by a seizure. But what surfaced was never predictable. Another of Penfield's patients, a jazz fan, had seizures that were precipitated by music, but also felt when they began as if he were experiencing a flashback. Successive stimulations of the same place on his brain prompted sensations of being back in the school washroom, on a street corner in South Bend, Indiana, and hearing an orchestration of *Guys and Dolls*.

Wilder Penfield had no doubt that he had discovered that the brain preserves all or nearly all past experiences "in astonishing detail," and that for some reason many of these are not retrievable by the normal processes of memory. It was a remarkable discovery, and one with a great deal of appeal, because it suggested that our past still lived on in the brain—in wonderful detail—waiting to be reawakened.

In the years since Penfield wrote about these patients, some objections have been raised to the claim that these stories represented real memories. Elizabeth Loftus, a psychologist at the University of Washington, has pointed out that only a very small percentage of Penfield's patients could be said to have experienced anything like a real memory while their brains were being stimulated, and that even in the very best examples there were reasons to be skeptical. Why, she asked, did M.M. feel as if she were in the lumber yard, when she herself admitted she had never been there? In another case a patient saw herself in childbirth, an out-of-body point of view that rarely appears in true memories. Loftus has argued that most of the cases which Penfield interpreted as reliving an experience were actually reconstructing it: seeing visions or hearing voices and rationalizing a context for them. This controversy may never be resolved, in part because there are just too few opportunities to explore the brain in this fashion. This is an invasive technique that can only be contemplated when surgery is already planned. Today, there are techniques for "seeing" the mind in action that hold more promise for much less risk.

Both the EEG and exploring the naked surface of the brain have largely given way to high-tech imaging techniques like positron emission tomography or PET. PET scanners make only indirect measurements of brain activity—they report local increases in the flow of blood, not changes in electrical activity of the brain cells themselves, but it is presumed that wherever there is a sudden increase in blood flow, there is heightened activity of brain cells. A volunteer whose brain is being scanned has a small amount of radioactive water injected into a vein. That water makes its way to the brain (via the bloodstream) in about a minute, and for the next several minutes wherever blood vessels expand in the brain, more blood (and more radioactive water) appear on the scene. At all times this radioactive water is emitting positrons, sub-atomic particles related to electrons. The PET scanner catches them leaving the brain and extrapolates their paths backwards to find the precise location that was their point of departure. The more positrons coming from any particular part of the brain, the brighter the image in the PET scan.

A PET scan can thereby produce images of the brain in action. Ask a volunteer to read a word, and bright spots appear in PET images of the visual areas of the brain; ask him to think of a verb that goes with that word and an area in the left frontal lobe just behind the eyeball turns bright.

Even PET scanning—as impressive as it is—might be superseded by a related technique called magnetic resonance imaging. Hemoglobin, the chemical in our blood that carries oxygen, can be made to behave like a little magnet, with its magnetic properties varying according to how much oxygen is actually on board. By putting a person in a magnetic field—nothing needs to be injected—scientists can identify places in the brain where oxygen-rich hemoglobin is present. Those are areas of high brain cell activity. Although sports fans might be familiar with MRI as a way of getting detailed and accurate pictures of the injuries to major-league baseball pitchers' arms, the newest adaptations go beyond PET scanning in revealing what's going on in the brain. PET provides snapshots; the new MRI creates the equivalent of movies of the brain at work. Researchers are falling all over themselves using both of these techniques to get pictures of any and every mental state they can imagine: what the brain looks like when it is sight-reading music; when it is improvising music; what goes on in the schizophrenic brain, the depressed brain, the teenage brain? The way the images correlate precisely with what the brain is supposed to be doing is as impressive as PET: a person being scanned thinks about a noun, and one area lights up; then s/he comes up with a verb to go with that noun, and a different brain area becomes active.

The new imaging technologies are a dramatic step forward from what was previously available. However, I can't help but feel that when this flood of research subsides and the experts sift through the millions of brain images that have been produced, the knowledge gained might not be quite as significant as it appears to be right now. A PET image shows that blood flow has suddenly increased in this or that part of the brain. Does that really bring us much closer to illustrating what individual brain cells are doing when the brain is "thinking," let alone some sort of biological comprehension of thoughts or desires? The ultimate goal of brain science is to get as close as possible to understanding human ideas and feelings in terms of brain cells and their electrochemical signalling systems, the ultimate hardware of the brain. Imaging alone is not likely to make that connection.

3
Crossing the Synapse

A single brain cell, or neuron, is a very powerful piece of communication machinery with a very low-tech appearance. Under the microscope a typical cell from the cerebral cortex looks like an ancient tree in dire need of pruning. The main trunk branches and then rebranches, over and over, mushrooming into a tangled mesh-work at the outermost reaches, all confined in a space fractions of a millimetre on each side. Cells like this are packed together throughout the cerebral cortex, and each one uses those branches to communicate with other cells. Ten thousand connections are made by each of a hundred billion cells; numbers like this provide a glimpse of how the brain is able to process complex information with the rapidity and intelligence that it does.

The outbursts of electricity seen in the brain with the EEG and the spots of colour on PET scanning images represent the activity of millions of neurons, or brain cells, all firing at the same time. It is their activity that underlies everything the brain does, from controlling a skater's take-off leg in a triple lutz to remembering the words to "She Loves You." The "firing" of a neuron is a brief event: a wave of electricity sweeps from one end of the cell to the other, stopping only when it reaches the tips of the branches. This burst of electricity remains confined to the interior of the cell; to convey the signal to the next cell in line, the neuron switches media from electricity to chemistry. The chemistry of neurons holds more fascination than their electricity, partly because it is easier to see in the mind's eye, partly because it provides a better way of understanding many of the puzzles of the brain and behaviour.

When the electrical impulse arrives at the end(s) of a neuron a strange thing happens. Hundreds of tiny sacs migrate to the inner surface of the cell membrane and fuse with it, discharging their contents to the outside world—in this case the space between brain cells. Each of these sacs contains several thousand molecules called neurotransmitters whose sole task in life is to carry neural messages from one brain cell to the next. Once released from the cell, they drift about aimlessly in the gap between neighbouring cells, but because this space is unimaginably small (about one fifty-thousandth of a millimetre), most of the transmitter molecules inevitably come into contact with the adjacent cell. Scattered across the surface of that cell are receptors whose shape mirrors that of the transmitters, and if the two come in contact, they momentarily bind together. This place where one brain cell talks to another is called the synapse.

There are tried-and-true analogies for this process—a key being fitted into a lock, a baseball in a glove—but none really does justice either to the subtlety of the binding or to the complexity of the events that ensue. For one thing, the receptors are much more intricate architecturally than gloves or even locks: each is a long chainlike molecule (assembled from a pool of many different kinds of links) folded upon itself so as to create a receptacle that is exclusively suited to the shape of the transmitter molecule. The receptors in turn are embedded in gate proteins, molecules with channels running through their cores which are forced open when a receptor grasps its

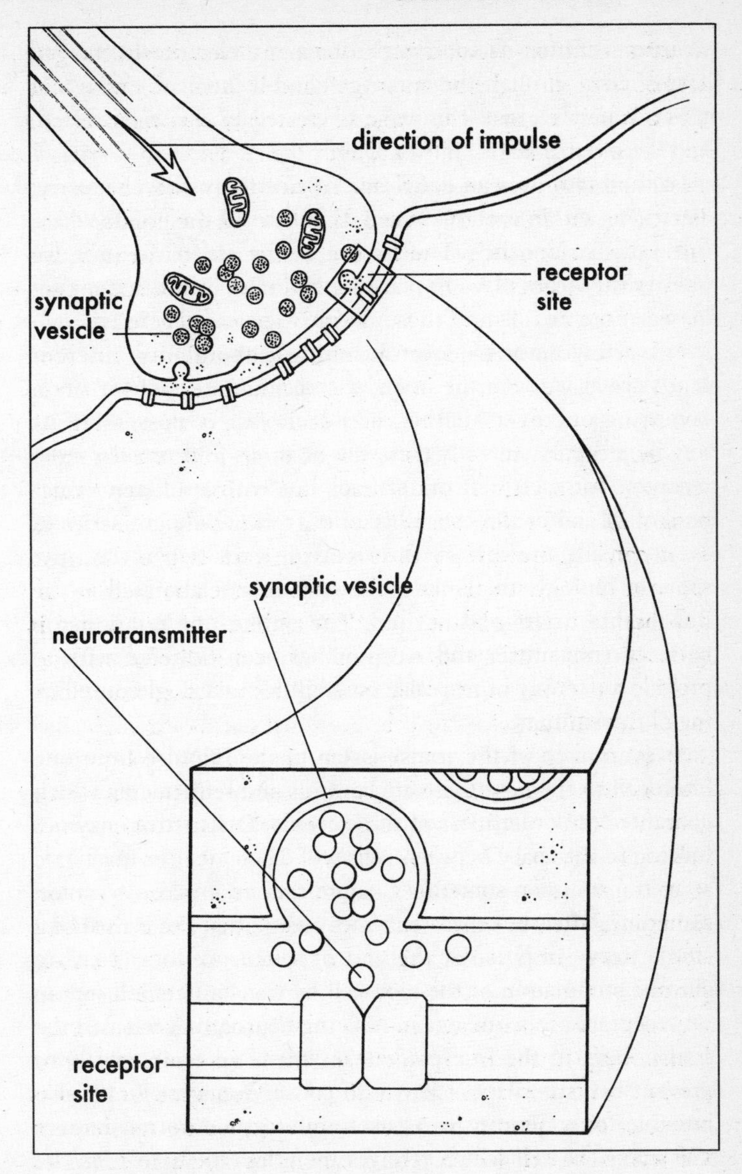

The arrival of a nerve impulse at the synapse
and the subsequent release of neurotransmitters.

neurotransmitter. As soon as that happens, electrically charged atoms move through the open gate, and if there are enough of those, a new electrical impulse is created in this second cell, and the process starts all over again.

So the brain uses an equal mix of electricity and chemistry: electricity within each brain cell, chemistry at the points where one talks to another. The nature of the electrical impulse doesn't vary from place to place in the brain, or for that matter between our brains and those of other species. There is, however, variety among neurotransmitters: about fifty different kinds are at work in the brain, with certain areas of the brain favouring one over another, and each variety possessing its unique receptor. Actually there can be three, four or even more receptors for each transmitter, each one with a different snugness of fit and/or the capability of triggering different series of events inside the cell on the receiving end. It is always the same in biology: the simplest of systems is elaborated to the imaginable limits and beyond. The one-to-one relationship between transmitter and receptor has been tinkered with to provide a panoply of response possibilities to a single outpouring of transmitters.

At every step of the transmission of the impulse from one brain cell to the next there are controls and refinements which guarantee the orderliness of these events. Destructive enzymes lurking in the space between cells will dismember or inactivate stray transmitters, sometimes before they've found a receptor, sometimes after, but these enzymes ensure that the arrival of a single nerve impulse at the end of a neuron doesn't create chronic stimulation of the next cell by transmitters left free to hit receptors again and again. Also the neuron that released the transmitters in the first place takes them up again, partly to prevent overstimulation, partly to conserve materials. It's also possible for a neuron to release inhibitory neurotransmitters that make the cell which receives them less likely to fire. The firing of a single brain cell will then depend on the balance between inhibition and excitation at any moment.

Remembering that the EEG records the ongoing electrical activity inside millions of neurons in the brain, imagine what must be happening in the gaps between those millions: every single electrical impulse generates frantic activity of transmitters flooding into the gap between cells, plugging into receptors, then being destroyed or taken up again. Each neuron is probably communicating in this way to thousands of others, and each can go through this entire cycle in a few thousandths of a second. Add the complication of fifty different neurotransmitters, many of which have multiple receptors, and you can see why so far it has been impossible to follow the chemical activity of the brain in the same way you can watch it electrically.

Nonetheless this picture of the micro-events between neurons has revolutionized the understanding of the brain by providing a simple model for answering the question, "What happens if something goes wrong here...or there...or over here?" It's a little like describing baseball solely in terms of the pitcher and the batter. The more questions you ask about this one fragment of the game, the more is revealed: "What happens if the ball is hit outside the foul lines, or straight up? Which is better when there is a runner at first, hitting the ball to the left or right side of the infield? Why is a right-handed batter likely to have more success against a left-handed pitcher?" The game can be built up from exploring the simple interaction of bat and ball. The same can be done by reducing the brain to receptors and transmitters: "What if there are too few, or too many receptors? What if foreign chemicals resembling transmitters arrive on the site and block the receptors? What if other chemicals prevent the transmitters from leaving the synapse so that they repeatedly stimulate the available receptors?" All these examples shed some light on how the brain works.

In the mid-1950s, anti-psychotic drugs were introduced for the treatment of schizophrenia, and while obviously they didn't eliminate the condition, they dramatically reduced the numbers of schizophrenics confined to mental institutions. The most effective drugs were those that blocked the receptors for a neurotransmitter called dopamine. That was the first real evidence that the hallucinations, delusions and disordered thoughts of schizophrenia might be the result of an excess of a brain chemical. As I am writing this, the latest thoughts on dopamine and schizophrenia are focused on just one of the several different kinds of dopamine receptors in the brain, and studies in Toronto have suggested that there is a six-fold increase in the number of these receptors in the brains of schizophrenics. Obviously more receptors present greater opportunities for a receptor-transmitter bond and therefore more frequent triggering of brain cells to fire. Admittedly it's still a considerable leap to connect the excessive activity of some dopamine-containing cells in some places in the brain to the strange behaviours and beliefs that accompany schizophrenia, and it's likely that those receptors are only part of the picture, but a connection of some kind can't be denied when a specific set of events at synapses can change a person's behaviour dramatically.

There is a curious link here to Parkinson's disease, one that can be readily explained at the receptor level. Parkinson's disease is predominantly a movement disorder: patients experience tremor, rigidity and have trouble initiating movements. The problem has been clearly identified as the gradual death of cells in a specific area in the middle of the brain. These cells extend branches to movement-regulating cells elsewhere, and employ the same dopamine as has been implicated in schizophrenia to convey these messages. Obviously as the population of transmitting cells shrinks, the amount of dopamine available to trigger the second set of cells into action is depleted. One of the most effective treatments is to give patients L-DOPA, a version of dopamine that can gain access to the brain from the bloodstream. From the receptor-transmitter point of view Parkinson's is a mirror-image of schizophrenia, and this is graphically demonstrated if treatments for either condition are not carefully monitored. An excessive dose of antipsychotic drugs given to schizophrenics may lower the effectiveness of dopamine so much that the symptoms of Parkinson's appear: tremor, slowness and dulled facial expressions. On the other hand, too much L-DOPA given to relieve the movement problems of a Parkinson's patient may make them temporarily psychotic. It is a classic example of how the same neurotransmitter molecule can play two completely different roles. In Parkinson's we see the effects of dopamine depletion in brain-cell circuits responsible for movement; in schizophrenia the same dopamine apparently hyperstimulates circuits which have their roots in the emotional areas of the brain.

The idea that receptors may come in slightly different shapes and sizes has provided the answer for a very different puzzle. In Rudyard Kipling's *The Jungle Book* the mongoose Rikki-Tikki-Tavi has no fear of the cobras in the garden, and Kipling first discards the old explanation that the mongoose eats a special herb to protect himself against the cobra's venom, then offers his own: "The victory is only a matter of quickness of eye and quickness of foot, and as no eye can follow the motion of a snake's head when it strikes, that makes things much more wonderful than any magic herb." Kipling is only half right: the mongoose does not rely on some magic herb, but neither does he rely only on his speed. Rikki-Tikki-Tavi has something "much more wonderful" going for him. The venom of the cobra is a receptor blocker, although the receptors it blocks are not on other neurons, but on muscle cells that receive transmitters from the nerves that command them to contract. Normally a nerve impulse travels to the end of a so-called "motor" nerve and triggers the release of the transmitter substance acetylcholine. If enough acetylcholine molecules plug into the receptors on the muscle cells, that cell will contract; taken together, nervous system commands to "jump" cause millions of muscle cells to contract simultaneously and you jump. Cobra venom blocks those receptors and paralyzes the muscles, including those controlling the movements of breathing. Death comes from respiratory failure.

Rikki-Tikki-Tavi has his ancient mongoose ancestors to thank for his ability to shrug off the effects of cobra venom. The mongoose has evolved receptors on its muscle cells that have a slightly different shape than the usual, with the result that the cobra venom molecule cannot bind to them. However, this change of shape is not dramatic enough to preclude the attachment of acetylcholine. So flooding the site with venom molecules has little or no effect; the mongoose's muscles still move and it kills the cobra. Ironically there is one other animal that has evolved the same strategy to achieve immunity to the toxin: the snake itself.

Another muscle receptor blocker is atropine, a chemical from the deadly nightshade plant which is similar in action to cobra venom. But it is much easier to control the effects of atropine, and in fact its alternate name, belladonna or "beautiful woman," was derived from one of its former uses. Women in ancient Rome and Egypt used to put drops of belladonna in their eyes, and when the poison blocked receptors on the muscles that control the iris, the women's pupils would dilate. This was done in the belief that dilated eyes made them more beautiful, and psychological studies in the 1970s at the University of Toronto showed that there was some justification for that belief. Apparently our pupils dilate if we happen to be looking at something that we desire: if you're hungry and you see a piece of chocolate cake your pupils dilate, but if you've just finished an enormous meal and you see the cake, your pupils don't change. Students asked to compare two identical pictures of a woman, one of which was doctored to enlarge her pupils, tended to pick that version over the unretouched one. To explain the effect of eyedrops on attractiveness, you have to add one more bit of psychology: a man looking deep into the eyes of a woman whose pupils are dilated will unconsciously realize that she appears to be interested in him, and without knowing why, will be attracted to her.

The human brain has more than fifty kinds of neurotransmitters, and they vary both in their molecular details and the effects they have on behaviour and thought. But other animals manage with far fewer transmitters, and the leech—or bloodsucker—is a perfect example. By the time you discover the leech clinging to your bare leg (half-full of your blood) it has already executed a beautifully choreographed set of feeding behaviours, every one of them made possible by the same neurotransmitter, serotonin. The ripples created by your entrance into the water alerted the leech to your presence and initiated vigorous swimming movements in your direction. When it came in contact with your skin it sought a warm place to bite, then ensured that your blood would continue flowing by injecting an anti-coagulant saliva into the bite as it was feeding.

Had you not interrupted the animal in the middle of dinner, it would eventually have stopped feeding anyway when its distended stomach signalled that it was full. The chemical serotonin involved in every one of these steps is also a vital transmitter in the human brain. It plays an important role in mood, a fact that has been underlined by the dramatic success of the antidepressant drug Prozac. Prozac's success has also embroiled it in controversy, because while it elevates mood in people who are clinically depressed, it has in some cases also helped those much less incapacitated, who could be said simply to be overwhelmed by the drudgery of everyday life. People who are excessively sensitive to criticism, lack self-esteem, fear rejection, or any combination of the above may respond to Prozac. The ethical dilemma is whether to prescribe Prozac for people who until now have relied on therapy rather than medication, if they have sought treatment at all. For some it conjures up images of a society more and more openly dependent on mood-altering drugs than it is now with nicotine, caffeine and alcohol.

Prozac interferes with the re-entry of serotonin into the cell that originally released it, and the transmitter's prolonged presence in the gap between nerve cells means that an individual serotonin molecule might interact with receptors two or three times for every nerve impulse, rather than once. It is as if the levels of serotonin have been boosted. (LSD and mescaline also have this effect because their shapes mimic that of serotonin.) But puzzles about Prozac remain. There are, amazingly enough, fourteen different serotonin receptors in the brain, many of them occupying their own territories in the brain. Presumably each uses serotonin for its own ends, which would explain why earlier serotonin-boosting drugs like the anti-depressive drug imipramine relieves depression but also causes side effects like dry mouth and abnormal heart rhythms. But how then does Prozac manage to avoid those side effects while targeting only certain behavioural problems? Why does it take several weeks for Prozac's effects to begin to be felt? The fact is that (at least at the time of writing) these questions cannot be answered because no one knows exactly what happens at the serotonin synapses when Prozac comes on the scene.*

I've outlined here what happens when neurotransmitters swarm across the space between two brain cells, but it should be admitted just how coarse this picture is. If you were to look in finer detail at the cell bearing the receptors, you'd see that the act of transmitters and receptors coming together triggers a welter of activity in countless chemical cycles, all of which refine the cell's response to that one synaptic event. There are still plenty of details of that event that aren't yet understood. That's just a moment in the life of one synapse, and when you realize that each neuron in the brain is keeping track of hundreds—if not thousands—of these at every moment, and that neurons are packed so tightly in the brain that you literally wouldn't be able to see through the thicket of their trunks and branches, then the challenge of truly understanding how the brain works becomes clear. And the real difficulty is to connect what we all know and experience as thoughts and feelings and

* Lab studies of leeches have shown that if you soak them in serotonin they will go through their whole feeding sequence with much greater enthusiasm. Imagine the unfortunate cottager knocking his entire summer's supply of Prozac into the water beside the dock, and leaping in after it only to be confronted with a lakeful of hyperactive serotonin-stimulated bloodsuckers.

mental images to these chemical and electrical events. It is collections of neurons that generate mental activity, but at what point does one become the other? How exactly do a few hundred thousand neurons or their neurotransmitters co-operate to create a memory of your grandmother's kitchen? We may never know, and it would be unlikely we would ever get there working from neurons up. Starting with the whole brain and analyzing its parts is probably more sensible, and in the last few years this approach has shown that there are surprises in the brain no matter where you look.

part two
NEGLECT

4
The Cathedral Square

The cathedral square, the Piazza del Duomo, is the living heart of the northern Italian city of Milan. The square is dominated at the east end by the great cathedral, begun in 1386 but not completed until 1965, when the last of five bronze doors was installed. The cathedral impresses by size and complexity if not by elegance: made entirely from white and pink Italian marble, it's one of the biggest churches in the world, capable of accommodating up to 40,000 people (although full houses are said to be rare). Tourist books play up the numbers: 135 marble spires, 2,245 marble statues, and inside, 52 eighty-foot pillars, each bearing a crown of statues. D.H. Lawrence called it "an imitation hedgehog of a cathedral," but others have been a little more generous: Bill Bryson, in *Neither Here Nor There* thought it was "quite splendid in a murky sort of way."

In front of the cathedral is an open space about the size of a football field, and in the middle stands a statue of King Victor Emmanuel urging his troops on in battle. All the buildings which border the square are known intimately to the Milanese,

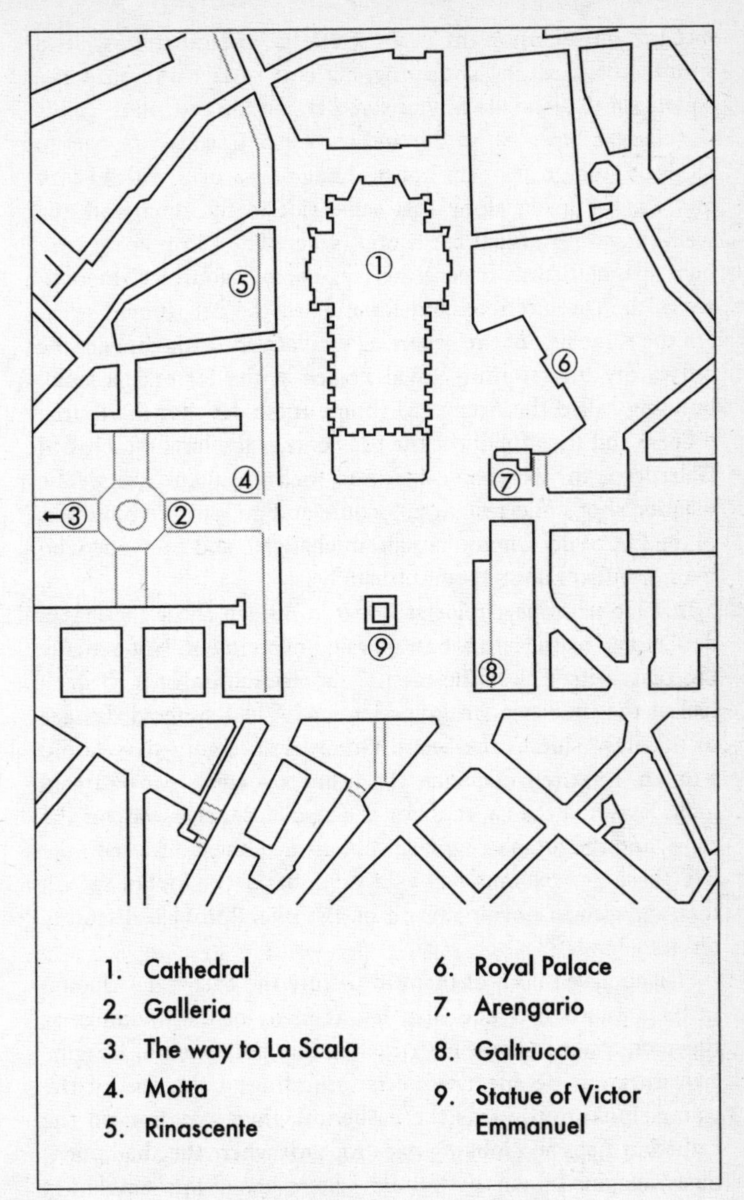

Cathedral square—aerial view.

but the most important is the Galleria, the great glassed-in nineteenth-century shopping mall. It opens out onto the square on the left side as you stand facing the cathedral, and is *the* place to meet, eat, drink and converse. It is also the way to La Scala, the great opera house. Imagine—a mall with history and class. Further along that same side of the square are the well-known restaurant/bar Motta (where you can prepare your own spaghetti dish from either regular or "gourmet" ingredients), the Duomo hotel and Rinascente, the department store. On the right side of the square as you face the cathedral are the outwardly unassuming Royal Palace at the far end, a small building called the Arengario (home these days for the tourist offices) and then finally at the near corner the long-established Galtrucco, for expensive made-to-measure men's suits. The tenants, shops and restaurants come and go, but the buildings of the Piazza del Duomo remain unchanged, and everyone who lives in Milan knows them intimately.

In 1978 neuropsychologist Edoardo Bisiach chose the Piazza del Duomo as the site for an unusual experiment. Not actually the piazza itself, but the mental representation of it. Bisiach had at the time two stroke patients who had suffered damage to the right side of the brain. One was an eighty-six-year-old woman, a retired manager, the other a seventy-two-year-old male lawyer. Because most of the stroke damage was on the right, and the brain's language centres are almost always on the left, these two patients had escaped aphasia, the loss of speech that is the most obvious result of a stroke. But what disability did they have?

Bisiach asked both of them to picture the Piazza del Duomo in their mind's eye, and then list as many of the buildings as they could remember. The trick was that they were to imagine two different perspectives: first standing at the end of the square looking towards the cathedral, then standing on the cathedral steps and looking back towards where they had previously imagined themselves to be. The results of this simple test cast a dramatic new light on one of the strangest phenomena in the neuropsychologist's case book: unilateral neglect.

The eighty-six-year-old retired manager, when imagining herself looking towards the cathedral, listed the Royal Palace, the set of stairs just in front of the Palace, the Arengario and further away (and not actually visible to someone standing in the square) the Via delle Ore. All of these are on the right side of the square as you look towards the cathedral. She named nothing on the left at all. But when asked to describe the vista from the cathedral steps, she came up with an entirely different list of buildings, including jewellers' shops, a shirt shop, Motta and the Via Dante—all of them on the right as you look from the cathedral steps back down the square.

The lawyer did the same. Imagining the square with the cathedral in front of him, he listed the cathedral itself, the corner of the Royal Palace, the Arengario, and Galtrucco (the men's wear shop at the extreme right of this perspective). After mentally switching to the cathedral steps and looking back, he listed a whole new set of buildings, streets, shops and restaurants, including Motta and Rinascente. Again these are all on the right side if you are standing in front of the cathedral.

This bias towards the right side of things (and lack of awareness of the left) is called neglect. It usually happens after damage to the right part of the brain, especially in the parietal lobe, above and behind the right ear. Neglect patients act as if the left side of space doesn't exist. They eat only from the right side of the plate. If the plate is rotated so that the remaining food is now on the right side, the patient usually claims that the food somehow mysteriously appeared out of nowhere. (They're right: for them the left side *is* nowhere.) They may read only the right-hand words in a newspaper column, or even the right side of each word, or shave only the right side of the face (which of course is on the right in the mirror). Neglect is most severe immediately after a stroke, usually fading as the weeks pass.

This little sketch of neglect suggests that these patients might just have a strange form of tunnel-vision, with the tunnel skewed to the left. But that is not true—it has been demonstrated over and over that damaged vision, although it might appear together with neglect as a result of the same stroke, doesn't necessarily lead to neglect, nor does neglect depend upon visual defects. Neglect is something beyond vision.

Edoardo Bisiach's experiment turned neglect into something much more dramatic—and curious—than it had been before. He showed that patients suffering from neglect have a bias towards the right, not just when they're looking at something, but even when they are *imagining* it. Neglect can't be dependent on vision, because their mind's-eye world is skewed to the right as well. The lawyer in this experiment was also asked to describe the office he had worked in for decades, again from two perspectives, coming in the door and sitting at his desk. Again he was only aware, or could only report, objects that were on his right.

This experiment also underlines just how profound this right-handed bias, or neglect of the left, is. The second description of the square and the office followed almost immediately upon the first, yet each contained a completely different set of landmarks. Yet at no time did these patients—or do most neglect patients—remark on the inconsistency or even absurdity of the situation. Where had the first set of buildings gone when each was reporting from the cathedral steps? Who knows? When prodded both were able to add to their list, even in one case mentioning a building on the left side, but fundamentally these two patients, although each obviously had a complete set of buildings in memory, were able to scan the right side of that mental layout only.

This is a little more complicated than it might seem at first glance, because a typical resident of Milan probably has in his head both a list of buildings in the piazza and their location relative to each other. These might be two independent stores of information or the list of buildings might be inseparable from the imagined map of the square; either way, both the layout and the memory are disturbed in neglect.

The idea that you can neglect half of the world around you is difficult to understand. You might think that the obvious therapy for such patients would be to encourage them to train themselves to look at the left side of the plate, or to work harder to imagine the left side of the square. But although neurologist and author Oliver Sacks described a neglect patient who did learn to turn all the way around to her right (close to 360 degrees) to view things on the left (and even requested a rotating wheelchair to make it easier), most of the time this isn't feasible. People who are aware they have a blind spot obscuring the left side do learn to move their head to compensate. But people with neglect are usually unaware they have a problem. They literally don't know what they're missing. They don't see their mental image of the Piazza del Duomo as full and clear on the right and somehow fuzzy and/or abbreviated on the left. What they imagine, or perceive, is what there is. It is not that they have lost the concept of "left"—the patients in Bisiach's experiment knew the left-to-right distribution of the buildings they were aware of—it is that there is much less "left" than you or I would see or imagine. You can't convince someone who doesn't believe there's any more "left" out there to turn farther in that direction. Can one who is not experiencing neglect envision this experience? Not directly, but I think it is possible to get a feel for this peculiar alteration of personal space by adopting a completely different point of view.

In a book called *Flatland*, published more than a hundred years ago, author Edwin Abbott described what life would be like in two dimensions. Flatlanders, an assortment of circles, triangles, squares and straight lines, live in a world with no vertical dimension, no up and down. They are entirely unaware—and cannot even conceive—of a third dimension. They glide about in a surface film of a world, identifying each other by the reflection of light from their sides: a circle presents an image that is bright in the centre and fades gradually away on both sides, a square may be uniformly bright if seen side-on, but will appear as a bright point if presented corner first.

The hero of the story, a Flatlander named A. Square, is confronted by a stranger from a three-dimensional world called Spaceland. The stranger is a sphere, something that is completely alien to A. Square. As long as the sphere hovers over Flatland, he remains invisible (though audible) to A. Square. A. Square can't look up, nor could he even understand what "looking up" would mean. When the sphere descends and pushes his way partly through Flatland, all that A. Square can see is a circle, the slice of the sphere that is at that moment in the plane of Flatland. To prove that he is more than just a circle, the sphere leaves Flatland by floating slowly upward. But to A. Square, this is no proof of an extra dimension, merely the demonstration that this particular circle was some kind of "extremely clever juggler" who had the unusual ability of being able to shrink to a point and even disappear completely, only to reappear unpredictably wherever he chose.

A. Square was completely incapable of imagining that this circle whose size changed with time could be a single intact three-dimensional object. A patient suffering from neglect is equally unaware of the full dimensions of space, and the cathedral square experiment suggests that they would have great difficulty remembering a time when they *were* aware of the spatial world in its entirety. Edoardo Bisiach's experiment established that neglect affects not just the scene a patient may be viewing, but also the mental representation of that scene in the brain. It's an experiment that has served as the theme for many variations, each of which has yielded new insights into neglect.

One of the most startling twists was reported in 1993, again by Italian researchers, this time at the University of Rome. Their patient, a stroke victim, showed no signs of neglect in any of the standard psychological tests. These include putting a mark in the exact centre of a horizontal line (neglect patients put their mark far to the right of centre, as if the left side of the line doesn't exist), or drawing a clock face (neglect patients either pack all twelve numbers onto the right side of the face or simply leave out the numbers 7 through 11.) Unlike this patient, Bisiach's two subjects had shown clear signs of neglect on such tests. This man apparently had a normal sense of the space he was viewing at any moment, yet when asked to describe from memory the layout of a famous square in Rome, the Piazza dei Cinquecento, he consistently missed buildings on the left, no matter which direction he imagined himself to be facing. (It's reassuring that most of these experiments have been done in Italy with Italian piazzas. You can imagine what subjects' testimony will be when and if these experiments are replicated in North America: "There's the McDonald's on the right just past the Taco Bell. The entrance to the mall is just behind the Wal-Mart with the six-decker parking lot behind that...")

Whereas Bisiach showed that neglect affects the mental landscape as well as the one being viewed, this man illustrates that the two can be separated, and that it is possible to be unable to pay attention to the left side of a mental scene, even though you might be perfectly capable of seeing, or attending to all of an actual scene you are viewing. This raises all kinds of interesting questions. If this man were actually standing in the Piazza dei Cinquecento, and looking at the buildings on all sides, then was asked to close his eyes and name them, could he do it? If he could, would a half-minute to clear his short-term or "working" memory mean that he would lose that recently remembered picture of the square and, having to rely on a mental image, revert to neglecting the left? What if he were standing at one end, with his eyes open, and was asked to imagine standing at the other end looking back at himself—what would he describe?

This particular case was even more interesting because the damage to the brain was not in the area that is almost always affected in neglect: high up on the right side in what's called the parietal lobe. This man had suffered damage to the right frontal lobe—much farther forward, and practically on the other side of the country as far as the brain is concerned. It's not clear what the significance of this is. There are also patients who exhibit the reverse set of symptoms: severe neglect on all the standard tests, but an unimpaired ability to list from memory the buildings around a square, or describe what can be seen when driving in either direction on the same road. The bottom line is that in neglect the mental landscape can be intact while the visual is impaired, or vice-versa.

It's difficult to know what, if anything, these patients think about their situation. Most of the time they seem unaware of it, like Flatlanders accepting the dimensions of the space they know. One German patient attributed his near-collisions with cars as he was out walking to the "rude behaviour of the Berliners." Another would go to the local bakery every day, which was a short walk and a right turn from his home. On coming out of the bakery he would often turn right again (because of his neglect), find he'd gone too far, turn back and walk right past his house, (now on his left), realize again he'd made a mistake, make an about-face and finally get home. He attributed his inability to walk directly home from the bakery to "absentmindedness," and came up with the classic neglect patient's view of the whole matter: "Your idea that I cannot see the whole space is one of the typical inventions of a scientist."

The more clinical reports of neglect patients you read, the more struck you are, not only by the strange symptoms but also by the listlessness of these patients. They just don't seem to care. Some of that could be attributed to a kind of shock reaction to the terrible blow of a stroke, but even when neglect patients recover from the immediate effects of the stroke, and the neglect fades, they rarely offer any profound glimpses into the condition. So we are left wondering what it is like to inhabit the distorted landscape of the neglect patient and what it might tell us about the brain.

5

Inner Maps of Outer Space

The details of the space that is underrepresented in the minds of neglect patients are now being filled in: some researchers are trying to define the vertical as well as horizontal dimensions, while others are trying to determine if neglect reaches all the way to the farthest corners of space that we can perceive. One patient in England, an avid darts player who suffered a severe stroke on the right side of his brain, seems to have developed neglect of the immediate space around him, but not of space a dart's throw away. When asked to indicate the centre of a line eighteen inches away, he showed an overwhelming tendency to mark it far off to the right of centre, one of the standard signs of neglect of the left. But as soon as this same test was moved away to a distance of eight feet, he did much better. He was both more accurate in using a light pointer to indicate the centre of a line marked on a white board and in throwing a dart at the centre of that same line. Interestingly he showed no signs of neglecting his own body, being able to touch his own left knee, calf, thigh, wrist and shoulder unhesitatingly when asked. Putting all this together suggests an onion-skin representation of space in the brain with, in this case, only the area immediately around the body (but not on the body) neglected.

One of the oddest things (about a very odd condition) is the predominance of neglect of the left, caused by stroke damage to the right side of the brain, especially in the parietal and temporal lobes beside and above the ear. It isn't surprising that damage to one side impairs the ability to scan the opposite side of space. Almost all incoming sensory information crosses to the opposite side as it arrives at the brain. Sensation from the big toe on your right foot arrives at the left side of the brain. This is also true of outgoing motor commands, and explains why damage to the left side of the brain often paralyzes the right arm and/or leg. But while it is not puzzling that damage to the right brain produces neglect of the left, it is unexpected that those are almost the only cases of neglect that appear. Why aren't there just as many instances of neglect of the right side caused by damage to the left side of the brain?

There have been two kinds of answer to this: one suggests that it's an illusion—that indeed there probably *are* all kinds of people who do neglect the right, but because their left-sided brain damage also disrupts language centres, they are just not capable of conversing and revealing their neglect. It goes unmentioned, and therefore unnoticed. One major study indeed found that there were many more cases of right-sided neglect than anyone had thought, but that study doesn't appear to have swayed many of the researchers interested in neglect. Most of them still accept that neglect of the left—meaning damage to the right side of the brain—is much more common than vice-versa, and have tried to accommodate this fact in their theories of what causes this strange condition.

One way to look at it is to picture one side of a map in the brain, some sort of internal depiction of the world, having been warped or even cut off by the brain injury. When the brain calls up a picture of the Piazza del Duomo, that picture is deformed. But that deformation could affect only the momentary display of that picture in your mind, not the permanently stored version of it; Bisiach's experiments proved that somewhere in their long-term memory these patients maintain an accurate right- and left-sided record of all the buildings around the square. Being unable to view the complete record would be like having a library of high-quality video tapes, but a defective replay machine. Or having a perfect set of slides, but a screen with a tattered and torn left side.

But if neglect were a loss of one side of an internal mental image or map, how could experimenters coax more details from patients by prodding them to list more buildings—as was done in a variation of the Bisiach experiment? Either their mental picture is damaged—and therefore inaccessible—or it isn't. That there's something more to this is clearly illustrated by a simple experiment. Patients with left neglect were asked to make a mark on a line at exactly the halfway point. Neglect patients usually make that mark to the right of centre because they are unaware of most of the left end of the line. In this experiment the patients marked the line far to the right as expected, but when a small number was printed just to the left of the line, two things happened. The patients were able to report what the number was, and they also did a much better job of bisecting the line. But put numbers at both ends, or worse, at the right end of the line only, the patients returned to showing no awareness of the left side of the line, and made their bisecting mark far to the right of the actual centre of the line.

It's hard to explain these results on the basis of a warped or damaged internal map of the world around the patient. How could the appearance of a single number at the left end of the line—in the middle of the lost side of space—improve the patient's awareness of a distorted map? You'd have to resort to desperate arguments along the lines that the map is somehow folded (or otherwise partially obstructed) as a result of the brain injury and that forcing patients to pay attention to something on the left, like a number, causes the folded corner of the map to be pulled back temporarily to reveal what's there. This is just too elaborate an analogy to be believable. There is a much simpler explanation for neglect: that it is a problem of attention. In other words, not a defect in the scene your mind's eye scans, but control of the mind's eye itself.*

One version of this argues that each hemisphere of the brain is biased towards scanning and drawing attention to things on the opposite side of the body—everything from near (parts of the body) to far (the horizon), but that the left hemisphere's power to do this far exceeds the right hemisphere's. The result would be that all of us, not just stroke patients, would have our attention pulled to the right because of the imbalance between the hemispheres.

There is some evidence that this is true: babies turn more often to the right than the left when startled, even if the sound that startles them comes from a point exactly in the middle. There have also been some unusual investigations of what people like or dislike in pictures: two studies, one American, one British, found that most people prefer paintings or photographs in which the dominant elements appeared on the right side, or appear to form a sequence leading to the right, or are actually moving towards the right—in other words, paintings or photos which carry your eyes to the right when you view them.

It's also commonly claimed that people lost in some featureless landscape like the Arctic or in a blinding storm walk great distances only to end up where they started. In the mid-1930s a New York psychiatrist named Paul Schilder argued that the majority of such people circle to the right. He added that when you are asked to close your eyes and hold out both your arms,

* Before discarding the idea of a damaged internal map completely, however, it's worth mentioning that follow-up experiments to the original cathedral square study illustrated that when neglect patients were asked to pay attention in their minds either to the left or right side only of the square, they did name more buildings, but many of them were placed on the wrong side. Was this another hint of a map gone wrong?

you usually hold the right arm higher. (He didn't differentiate between right- and left-handers). Schilder saw these as examples that we all neglect the left to a certain extent. Today there are some neuro-experts who argue that all of this evidence—the picture preferences, babies head-turning and even the anecdotal reports—illustrates that the left hemisphere's bias towards the right is slightly stronger than the right hemisphere's similar tendency to the left, and can in some circumstances prevail.

If this is all true and there is an injury to the weaker, right side of the brain, the left is suddenly freer to exert its strong rightward pull on attention and movement. As a result, you have a patient who looks to the right, moves to the right and is only aware of the right.

A different attention scenario supposes that the right hemisphere of the brain is normally responsible for paying attention to both sides of space, while the left brain monitors only the right side. If this were true and the left hemisphere were damaged, both sides of space would still be monitored by the right hemisphere. (This would account for the much smaller number of people with neglect of the right.) But injury to the right side would leave the left of things unattended.

One slight twist on this version suggests that the problem isn't so much the tendency to pay attention to the right side, but the inability to shift attention after it's already been directed to the right. In other words, once fixated on the right, a patient with neglect, while not incapable of switching to the other side, has a lot of trouble doing it. The kinds of experiments that suggest that the disengagement of attention is the real problem are usually set up so that patients watch a screen and try to pay attention to brief and sudden appearances of simple objects (a square or an asterisk) popping up on either side. It is generally true that when the action is on the right, patients with neglect lose all awareness of the left. But if the only signal on the screen comes from the left, they are able to pay attention to it. The experiment which showed that patients are much better at dividing lines in half if a number appeared to the left of the end of the line is a perfect example. Such experiments fit the idea of a searchlight of attention that tends to swing automatically to the right, and will only move left if forced.

In the case of neglect, as with so many areas of brain science, imaging of the brain at work has painted a whole new picture of what is going on. One study published by Maurizio Corbetta and his colleagues at the Washington University School of Medicine early in 1993 revealed that the two hemispheres are indeed unequal when it comes to shifting attention from one target object to another. In this experiment volunteers sat facing a screen displaying an array of tiny boxes, ten in a row running from left to right with an eleventh above the line right in the middle. Participants fixed their gaze on a set of cross-hairs in this eleventh box, and without shifting their eyes, tried to pay attention as a flashing asterisk moved from box to box below. Most of the time, the asterisk followed a predictable pattern, moving steadily to the right or left with each appearance, then cycling back and starting again. At the same time, the volunteers' brains were being imaged by PET, which creates a picture of those places in the brain that are, at any moment, the sites of sudden increases in blood flow. It's assumed that when a local patch of brain cells become active, they require more blood. A "hot spot" on a PET scan is presumed to be a place where the brain is very active.

The results were startling. When the asterisk was flashing on and off in boxes on the left side of the line, the right side of the brain, as expected, was most active. There was a trace of activity in the left brain, but clearly the right brain was in control when attention was directed to the left. What was surprising was that the right brain was also hard at work when the asterisk was popping up on the right side, at a time when the left brain should have been directing shifts of attention from one box to the next.

Not only is the right hemisphere more involved than the left in shifting attention from one place to the next, it has more space devoted to the task. Two distinct areas towards the back of the brain on the right showed up in these PET images, one paying attention to the left, the other the right. In the same areas on the left side of the brain there was only one weakly reacting site that turned on regardless of where the shifting asterisk was. The imagery provides strong support for the conjecture that the right side of the brain is in control when it comes to attention, at least the kind of attention demanded by this experiment. This in turn creates an imbalance in the surveillance of space around you: paying attention to the left is almost exclusively under the control of the right side of the brain, but attention to the right side is shared by both right and left.

These images make perfect sense of the much larger numbers of stroke victims with neglect of the left. Damage to the right hemisphere can easily knock out the large and specialized area that is the *only* part of the brain paying attention to events on the left side. But equivalent damage to the left hemisphere, no matter how severe, would leave untouched the right hemisphere's own centre for directing attention to the right.

There is one nagging detail that demands to be taken into account when it comes time to theorize about neglect. It is the discovery that syringing ice-cold water directly on the left eardrum of a patient with neglect of the left reduces the problem dramatically, if only for a few moments. It is odd that these patients are untroubled by this treatment—for most people ice-cold water on the eardrum is excruciatingly painful. It is doubly odd that chilling the eardrum could enable someone to revisit, however briefly, the fullness of space around them. The usual explanation is that the shock of the water reactivates the hemisphere (in this case the right) which has lost the ability to direct attention to the "missing" side of space. As is typical with neglect, as the effect of the ice-water fades and neglect returns, the patient's memory of having experienced a wider panorama of space only moments before fades too.

If neglect results from an inability to direct attention to the left, or putting it another way, to disengage attention from the right, it might explain why people who suffer from neglect are usually incapable of explaining what it is like. How can you describe something you cannot direct your own attention to? If instead neglect were the result of a distorted mental map, you might expect patients to be able to cast their attentional searchlight on that distortion and describe it, but by and large that doesn't happen. There are some exceptions: in the 1950s William Battersby at the Mount Sinai Hospital in New York reported a case where a man, upon recovering from neglect, remembered how badly he had done previously on a set of tests:

> "The last time I read these things I would start at the left and jump all the way over to the right."
> "Why did you do that, Mr. B.?"
> "I don't know, what with the left side being so bad and all, I just didn't want to look that way."

A much more recent patient has also provided a tantalizing glimpse of the missing parts of space. A left-handed English gentleman farmer suffered two strokes in late 1981 and early 1982 which left him with, among other things, a peculiar inability to spell. When asked to spell aloud (his writing was too severely disturbed by the brain injuries) three-, four- or five-letter words, he chose to spell them backwards, apparently deliberately, and he made many more errors on the first letters of the word than the last. Thinking that this might be a memory problem—that by the time he got to the first letters he had forgotten what he was spelling—the investigators had him spell similar sets of words both frontwards and backwards. Again most of his mistakes were with the first one or two letters of these words. This is exactly the opposite of the spelling mistakes children tend to make. Doreen Baxter and Elizabeth Warrington, the psychologists who worked with this man, argue that this was a unique result of neglect: that this man was spelling from an "inner screen," but was unable to identify the left side of words (the first one or two letters) on that screen. He himself made the comment that trying to spell was "like reading off an image in which the letters on the right side were clearer than those on the left." This is one of the rare recorded comments on what the experience of neglect might be like.

This man is worth noting for another reason: if his neglect is supposed to be a result of the inability to focus attention on the left, or disengage from the right, why then was he able to "see" the fuzzy beginnings of words, and even to be able to judge how many letters at the beginnings of words he was getting wrong when he transcribed them? In this case the neglect applies to the words themselves, not the so-called inner screen on which he was reading them. So is it space, or is it attention, or both? The question just can't be answered yet.

6
THE BURNING HOUSE

In 1988 the English neuropsychologists John Marshall and Peter Halligan reported an extraordinary set of experiments involving a single patient with neglect. By showing her simple line drawings of a house, they revealed that someone who seems to be completely unaware of one side of her world may in fact know something about it, and yet be completely unaware of that knowledge.

The patient was a forty-nine-year-old woman who had had a brain haemorrhage. The result was a classic case of neglect: when copying drawings, she left out the left side; when asked to read words, she'd either leave the left end off ("simile" became "mile"), or reinvent it ("façade" changed to "arcade"). In the crucial experiment she was shown two cards one above the other, each of which had on it a simple kindergarten-style picture of a house, with nothing more than a roof, a chimney, five windows and a door. The two houses were exactly the same, drawn in black and white, except that the left side of one of them was enveloped in bright red flames. She was asked to describe the drawings: "A house," she said. "Are the two houses the same or different?" "They are the same." "Anything wrong with either card?" "No."

Then seventeen different times, first with the burning-house card above, then below the other card, she was asked the seemingly ridiculous question, "Which house would you prefer to live in?" Even *she* thought the whole exercise was silly, especially since to her the houses were identical, or so she said. But the results were far from silly—out of the seventeen trials, she chose the house without the flames *fourteen* times. Yet all this time she claimed to notice no difference whatsoever between the two houses. When Halligan and Marshall then showed her new cards depicting the same house but with flames enveloping the right side, she chose the intact house every time, finally commenting in exasperation, "I hope there's a point to all this."

There was a point. The most reasonable way to explain why someone would choose one house consistently over another—all the time maintaining they were identical—is to surmise that subconsciously she was aware that one of the houses was less desirable than the other. In concrete terms, her brain had noticed the flames, interpreted what they meant, and was guiding her decision-making. Yet she had no idea any such process was involved.

An elaboration of this experiment revealed a little of what might be going on in someone's mind in situations like this. Edoardo Bisiach (of the cathedral square experiment) tried the same sort of experiment, but added some new objects to the burning houses, including two wineglasses, one intact, the other with a piece broken off on the left side of the rim, and two flowerpots, one with flowers arranged on the left side of the pot, one without. When Bisiach showed these drawings to patients with neglect, he found to his amazement that most of his patients could trace the outlines of these drawings with a finger with great precision, and yet all the while fail to notice the flowers or the chipped edge through which their finger passed. In the case of the house, the outline of the roof and upper wall on the left side was obscured by the flames, yet these neglect patients would move their fingers straight along where the roof and wall would have been. Never did they notice anything odd. Bisiach could only guess that the

activity of tracing was, when it came to the left side of these drawings, *absent-minded.**

But even the tracing isn't as significant as something else that the experimenters noticed with this set of patients. In the case of the two houses, some of them fabricated the most extraordinary reasons for choosing one house over the other. (Oddly enough, a couple of Bisiach's patients actually preferred the burning house, but even so, like Halligan and Marshall's patient, they asserted that the two were identical.) Some of these patients then offered the following reasons for their choice of house. One said it was "more roomy"; another claimed it was "more spacious, *especially in the attic*," and "had a better layout." This latter patient also preferred the unbroken wine glass, saying it was "more capacious," and then having mistaken the flowerpots for cooking pots, chose the pot with the flowers each time because it was "larger and better suited for cooking." He finally saw the flowers when a mirror-image version of the pair (with the flowers now on the right-hand side of the pot) was shown him. "Nice daisies," he said, all the time maintaining that these were cooking pots. And when asked why cooking pots would have daisies in them, he replied poignantly, "A gentle hand put them in."

These are doubly amazing reports. To imagine that someone could trace with a fingertip the outlines of a house, and in the process move through a mass of flames erupting from a window, and yet have no awareness of those flames is startling and underlines the fact that any theory of neglect must necessarily reach beyond the (relatively) simple phenomenon of inattention to one side of space to include explanations of how the

* The most stunning example of this to date has been provided by psychologist Andrew Young in England. He showed a patient a hybrid picture comprised of half a glass bowl on the right apparently joined to half a football on the left. The two halves were scaled so that the outline from bowl to football and back to bowl was continuous. Young's patient claimed the object was a bowl, period, and proceeded to prove it by tracing the outline of the rim of the bowl in a complete circle. The outline followed the rim of the half-bowl, then continued smoothly around the football back to the bowl.

patients in this experiment can be watching carefully what they are doing yet apparently not be able to see. In some of these cases the flames were visible to the right of the finger doing the tracing, yet the patient betrayed no notice of them.

The situation is even more mysterious when it comes to the patients claiming that they chose the unbroken wineglass because it was "more capacious," or the house without the flames because it was roomier, "especially in the attic." These explanations are called confabulations—they are "honest lies," made-up stories to rationalize a person's choice of one object consistently over another, all the while testifying that there is absolutely no difference between the two. These stories take the phenomenon to a new level. They shed light not just on how unaware of their left side neglect patients are, but also on how they deal with that lack of awareness. Such experiments also suggest that our brains can take in certain kinds of information and use it to direct our actions without our ever being aware that such information exists. That doesn't surprise brain scientists—there are other pieces of evidence that this happens—but these are particularly vivid examples. Why do patients suffering from neglect tell these stories, and where do the stories come from?

There are several possible answers, and the simplest might be that after allowing for the effects of the injury to the brain, these people aren't much different from the rest of us. There is plenty of evidence that in everyday life most people are unable to identify correctly the influences that prompt them to make decisions, especially when those decisions involve choices not dissimilar from those facing these patients. In the 1970s the American psychologists Timothy De Camp Wilson and Richard Nisbet performed a series of experiments designed to discover whether or not people know why they make certain choices. Can any of us identify the influences that prompt our decisions? The answer is usually no. In one experiment volunteers were given thirty seconds to memorize eight pairs of words (like jelly-purple, ocean-moon and pizza-olive) then they were tested for their retention of those pairs. In actuality, however, the purpose of the experiment was to see if the word pairs would influence the subjects' later responses to a word-association test. For instance, they'd be given the words "fruit," "brand of detergent" and "foreign country" in the anticipation that the word pairs mentioned above would prompt the answers "grapes," "Tide" and "Italy." And they did. Those expected answers were mentioned twice as often by subjects who had seen those particular word pairs than by control subjects who hadn't. So there was an unambiguous effect of the memorized word pairs on the answers in the word-association test. But you'd never know it from asking the people who took part. They were asked to list all the possible factors that might have influenced their word-association answers. They came up with twelve, ranging from distinctive features of the objects named through personal affection for the object to recent experience. Last on the list? The word cues, the one thing that had an obvious and potent influence on the answers these subjects gave.

Another Nisbet and Wilson experiment took place in a shopping centre. Outside a bargain store they set up a card-table with four identical pairs of panty hose arranged in a row, with a sign above reading: "Institute for Social Research—Consumer Evaluation Survey—Which is the Best Quality?" Fifty-two people participated, and although two suspected the stockings were all the same, the rest had no trouble choosing what they thought was the superior member of the four identical pairs. Position seemed to be the most important factor (the further right the panty hose, the more votes it received), but the participants claimed their decisions had been based on "knit, weave, sheerness, elasticity or workmanship."

Nisbet and Wilson conducted a variety of studies like these, and concluded that not only are most people unaware of the factors that are truly influencing their decisions—they just don't know—but they are also more than willing to identify a set of bogus factors—they tell more than they know. In many cases, including the examples I've cited here, the reasons given are reasonable. They are indeed the factors you'd expect to influence people, like a recent experience with grapes prompting the association with the word "fruit" or the sheerness of panty hose making them superior. Plausible but in these cases irrelevant.

If it's true that under normal circumstances we are often unable to identify the real factors that influence our choices, and instead rely on the most plausible, could that not also be true of these neglect patients? Obviously they start with what you would think would be a profound handicap. At the conscious level their brain damage leaves them unaware of the flames leaping from the house or the chipped rim of the wineglass, but the consistency of choices suggests there is probably some subconscious registration of the oddities in these pictures. But because the flames or the chipped rim are not available to the patient's consciousness, they then do what anybody might do in the circumstances: explain their preference for one glass or house over the other on the basis of the plausible factors—the "layout" of the house or the volume of the glass.

This may be part of the explanation for these experiments with the neglect patients, and if so it does underline how neglect is a peculiar combination of the strange and the familiar. None of us can really imagine being unable to experience the left side of things. On the other hand, judging from the Nisbet and Wilson experiments, we are all simultaneously incapable of identifying the reasons why we make choices and eager to explain those choices nonetheless. And we are certain that our rationalizations are correct.

That urge to explain may be traceable to the left side of the brain. Michael Gazzaniga of the Center for Neuroscience at the University of California at Davis has suggested that the left cerebral hemisphere contains an "interpreter," a centre whose job it is to make sense of contradictory or unexplainable information. This is an idea that has come out of the extensive research identifying differences between the right and left brain.

The idea that the two sides of the brain "think" differently had its origin in the examination of people who had brain surgery for epilepsy. In the 1950s surgeons in the United States revived a little-used surgical procedure called commissurotomy as a last resort for the treatment of severe cases of epilepsy. In cases where drugs were ineffective, and the seizures, although starting locally, spread rapidly throughout the brain, surgeons severed the main connection between the right and left hemispheres, the corpus callosum. The hoped-for result was that the seizures would at least then remain localized in one half of the brain. The unanticipated by-product was the discovery that people with these so-called "split brains" acted in some circumstances as if they had two independent brains or minds. Each hemisphere could be prompted to act on its own.

The flood of research that followed that discovery has had a huge impact on the understanding of how the brain works, even after subtracting the popular hype surrounding it. Michael Gazzaniga is one of the pioneers of experiments with split-brain patients, and he has never hesitated to paint the ideas gained from these experiments with bold strokes. It was he who first showed that careful experiments with split-brain patients revealed that in some situations they would rationalize in much the same way as the neglect patients do.

These experiments require sending different information to each brain hemisphere, something that is surprisingly easy to do. If you are seated directly in front of a screen with your eyes fixed straight ahead, a brief flash of an image off to the left will appear only in the extreme left corner of your visual space and will be sent only to the right hemisphere. Normally information about that image would then be shared instantly with the left hemisphere via the corpus callosum, but of course in split-brain patients that doesn't happen. In the same way images appearing for a split second off to the right are only "seen" by the left hemisphere.

This has led to some surprising, and even funny experiments, at least by psychology lab standards of humour. One described by Gazzaniga in the late seventies involved a patient who saw a picture of a chicken claw with the left hemisphere and a snowy scene with the right. Gazzaniga then used headphones to instruct the patient, or rather, each of his hemispheres in turn, to pick out a picture that matched the one already seen. This instruction fed into the right ear goes more or less directly to the left hemisphere, which has no trouble comprehending what is being asked because in most people it is the hemisphere containing the speech centres. It's also possible to send (through the left ear) the same simple instructions to the right hemisphere—its lack of language centres renders it incapable of giving a verbal answer, but it can still comprehend the spoken word. In this case the patient's left hand (controlled by the right hemisphere) chose a shovel to go with the snowy scene, and the left-hemisphere–right-hand combination picked out a chicken to go with the claw. But of course, neither hemisphere knew what picture the other had seen, so each was in the odd position of watching both hands pick out a picture, only one of which made sense.

But ignorance didn't prevent the brain from offering an explanation. The patient was asked why he picked the pictures he did. The only hemisphere with a voice is the left, but that hemisphere was completely unaware of the snowy scene, and therefore should have been at a loss to know why the left hand picked up the shovel. Undeterred, the patient (or rather his left hemisphere) answered, "Oh that's simple. The chicken claw goes with the chicken, and you need a shovel to clean out the chicken shed."

A plausible answer but one that had nothing to do with what had really happened. But this is not an uncommon tactic by the left hemisphere: Gazzaniga tells other stories, of how when the word "walk" is transmitted to the right hemisphere, a few seconds pass and then the patient stands up and starts to leave the room. When asked why, he (his left hemisphere) says, "I'm thirsty, I just thought I'd go down the hall and get a Coke."

This is where Gazzaniga's interpreter comes in. He argues

that the left hemisphere contains a centre that makes sense of, and then explains verbally, actions, moods or thoughts that it knows nothing about. There are hundreds of stories from the studies of split-brain patients like the one about the chicken shed and the shovel, and Gazzaniga thinks that this interpretive role is played by the left hemisphere even in normal, unsplit brains.

Supposedly our brains need the interpreter to combine the stream of disparate bits of information from a variety of brain centres with different specialties into one seamless conscious experience. When you jump up and say hello at the sight of someone you know coming into the room, you aren't aware that separate language, face-recognition, emotional and memory centres are all active at the same time. Michael Gazzaniga would say that's thanks to your left-brain interpreter. He has seen how the left hemisphere in split-brain patients will react to a change of mood initiated in the right hemisphere. No matter what the mood change, positive or negative, the interpreter will attribute the change to factors that have absolutely nothing to do with it, like the people present, the place, the time of day, the last meal, it doesn't matter. The interpreter must explain the mood change—that's its job.

The explanations of neglect patients for their choice of pictures sound suspiciously like the work of a left-brain interpreter. They all make the same kind of superficial sense and they are all statements about events the brain is apparently unaware of. Lack of awareness is the key: in split brains, the left and right sides of the brain no longer communicate, so the left hemisphere doesn't know what the right has done. In neglect, the left hemisphere is apparently unaware of features, like flames leaping from a window, which prompt the choice of the other house. The patient initially states, using the speech abilities of that same left hemisphere, that the pictures are no different, but is then forced into the imaginary world of confabulation.

Is it only in these (fortunately) rare cases of brain injury and/or surgery that we can see the actions of the left-brain interpreter? Michael Gazzaniga suggests that the interpreter is active all the time, synthesizing and making sense of different kinds of information from several independent brain centres, and the rationales offered by perfectly normal people for choosing a pair of panty hose as the best of an identical lot would seem to back him up. One of Gazzaniga's favourite examples is the establishment of a phobia. In his view someone could experience a panic attack that is entirely the result of a spontaneous change in brain chemistry, but the left-brain interpreter, in an effort to make sense of the intense emotional reaction, would then associate these unpleasant feelings with the surroundings of the moment. These might be a restaurant, a crowd, an elevator, whatever. The fear of these situations could then long outlast the actual attacks that they are used to explaining. Could we ever catch our own interpreter in the act? It's unlikely because the information with which it works is by definition, subconscious. This is just as well, because if this theory is correct (and there are brain scientists out there who have their doubts), the interpreter makes coherent a lot of conflicting, confusing and mysterious information that we are probably better off not knowing. It is the story the interpreter creates that we're aware of, not the events that prompted it to tell that story. It's only when we as outside observers are in the privileged position of knowing what the left brain doesn't, as in the cases of split-brains or neglect, that we can immediately recognize a left-hemisphere fable for what it is.

To recap: the brain maintains a map of the space around us, a dynamic three-dimensional virtual-reality-in-the-brain representation of space, and that map likely needs to be illuminated by the mental searchlight, the focus of attention that must be directed to any part of the scene desired before it comes alive. Although in some cases it appears as if the map itself can be damaged, it is also true that attention can be redirected to the left side, if only for a few moments, suggesting that neglect is the combination of an intact map with a defective searchlight.

(It's hard to imagine the other side of that coin, a patient who was perfectly capable of directing attention to any point in space, but with no map to point to.)

And what does the case of the burning house tell us specifically? It reminds us how all brains, damaged or not, work to provide their owners with a sensible conscious experience no matter how difficult that might be. Neglect patients are unaware of a large portion of the world around them. They are also unaware that they are unaware. And yet, doubly handicapped, when asked to explain their own thoughts they are as capable of resorting to the same kind of reasonable guessing as the rest of us. It seems odd that a man would think that a chipped wineglass was essentially the same as an unbroken one, but given that omission from his conscious store of information, he then does what we would all do: justify his preference for one over the other on the basis of plausible, but nonexistent differences. It is this resourcefulness, the unstoppable will of the brain to explain, justify and make sense of things no matter how impoverished the available information, that makes these cases both fascinating and pathetic.

part three
THE HOMUNCULUS

7
A Cup of Coffee

The next time you reach absent-mindedly for your morning cup of coffee, take a moment to reflect on just how you do it. It will require paying attention to an action so routine and well-practised that you have never before even been aware of it. But if you think that implies that such an action is therefore uncomplicated and without mystery, you are in for a surprise.

If you are like me, you begin with the briefest of glances in the general direction of the cup, calling on the memory of the last sip to guide your gaze and confirm that the cup is indeed where you expect it to be. At almost the same time, your arm will begin to extend, and that extension continues smoothly even after you've stopped looking at the cup. Contact with the handle can be made entirely by touch, although you may need to make a last-second visual check to ensure that your fingers thread through properly. Bringing the cup to your lips can be done entirely by feel: you can read, converse or just stare off into space while doing it without missing a beat or spilling a drop.

This sequence is possible only because your brain is busy creating, storing and merging a variety of mental maps and calculating the correct movements based on them. These include both maps of your body and limbs and their relative positions, and maps of the space around you. You are entirely unaware that this is going on, but it is possible to imagine life without them. What would it be like to reach for the cup on the table next to you if your brain was incapable of maintaining a record of the relative position of your hand and the cup? Guiding this simple act by vision alone would be a strange experience: immediately upon feeling the urge to take a sip, the brain stem (the part of the brain just above the spinal cord that takes care of simple eye movements) would flip the eyes into scanning mode, searching the neighbourhood of the right side of the body (for right-handers) looking for the arm, then the hand. (This is assuming you already know where the cup is—a problem in itself.) Once the hand was set in motion roughly in the direction of the cup, the eyes would be scanning frantically back and forth to ensure that the hand really was getting closer and closer to the handle, making constant corrections until finally the fingers made contact.

Once contact was made you could bring the cup to your lips in the normal fashion. Or could you? How would you know exactly where your lips were relative to the hand holding the cup unless you were juggling maps of both in your mind? In the absence of mental maps, what is now just one of hundreds of routine operations your brain handles in a day would become laborious and erratic.

While the experts acknowledge that mental maps are crucial for getting things done efficiently, there is great controversy over just what a brain map is like. Because the brain is computerlike, and because no one these days believes there's a little theatre with a Cineplex-sized screen in it buried deep under the cerebral cortex, most neuroscientists doubt that mental maps are displayed in the brain as they are in an atlas. When you call to mind a map of Canada, it's unlikely—most such experts would say—that an array of brain cells near your right ear representing Newfoundland suddenly starts firing, together with cells all the way across the cerebral cortex to Vancouver Island. (If it were true of course, PET scanning images of the brains of Torontonians might reveal what everyone west of Thunder Bay has always suspected: that their maps are curiously disoriented so as to centre their city on the map. Any map.) Today the feeling is that the information about space necessary for performing acts like reaching for your cup of coffee might be encoded or recorded in some other way. For example, you can find any place on a world map as long as you know its latitude and longitude—you don't require a mental image of its location. It's conceivable that the information about the location of objects around you could be recorded in some similar sort of code that would permit reaching for the object without ever having created a mental picture of it or its background. This isn't to say you can't close your eyes and picture it—of course you can—but that might be a completely different (and rarely necessary) process.

Reaching for the cup of coffee requires the brain to relate the location of the cup to the position of that hand and arm, and then to use that information to initiate appropriate movements. The first step, the location of the cup, depends on establishing a map of the space immediately surrounding you.

Scientists who are trying to understand how the brain envisages this space for the purposes of lifting a cup or playing the piano call this "grasping space." On the other hand, some social psychologists are just as interested in this nearest of spaces for completely different reasons. Their concern is not so much what you do with it as your feelings of ownership over it, and they call it "personal space." It's the area immediately next to your body which others may enter only by invitation. Many studies have established the dimensions of personal space (slightly larger in front than behind, ending just above the head) and have documented the discomfort we feel when that space is violated. Theorists of personal space have expended plenty of energy trying to explain how it is that we can calmly stand cheek-to-cheek with strangers in a subway car but would scream if one of them were to do the same in the middle of an office—while one's tolerance of invasions of personal space can be fitted to the situation, it's striking how discomfiting an inappropriate or untimely invasion can be. The fact that different cultures seem to establish personal spaces of different dimensions and that the amount of space a person is willing to share may expand or contract depending on the circumstances suggests that this space is not a fundamental construct of perception, but is a learned, and therefore much more malleable kind of "space."

On the other hand, the brain maintains a different mental model of exactly the same physical space, a model that allows it to direct the eyes, hands and body to operate within that space and which unlike "personal space" has no emotional content. However, neuropsychologist Otto-Joachim Grüsser of the Free University of Berlin emphasizes that this is only one of several spaces of different size around us (remember the patient in Chapter 5 who neglected the space a darts-throw away much less than he did the space within his reach.) Beyond the immediate space around you is an area that could be considered as the space within which you can operate comfortably. Grüsser claims that the familiar feeling of anxiety that overtakes you if you try walking across a schoolyard with your eyes closed marks the borders of this space: there comes a point after about ten to fifteen steps when you feel that you must open your eyes to get your bearings, even though you know perfectly well there is no danger in front of you. For the blind, hearing takes the place of sight for reconnoitring this near space.

Beyond near space, from about fifteen paces out all the way to the horizon and the sky, there is yet another space, one that differs from the two smaller-scale versions by being warped. The sky does not appear to us to be a perfectly spherical dome, but a flattened one: the horizon seems farther away than the point on the sky directly above. This has been shown by perceptual tests, but is easily experienced with the full moon, in the famous "moon illusion." The horizon moon seems much larger than the overhead moon, even though the simple test of holding an aspirin tablet at arm's length reveals that the moon is the same size whatever its elevation, and can just be covered by the pill.

The key to the illusion is our perception of the distance to the moon. Any object at the horizon seems further away than the same object overhead, because our visual "far-distant" space has a low ceiling. The brain then has to resolve the following conflict: when the moon is at the horizon it seems farther away than when overhead, but the image it makes on the retina of the eye is exactly the same size. The only way to make sense of this apparent contradiction is to assume that the moon must be considerably larger when it is near the horizon. A second psychological step is then taken, making the moon that is larger by reason, larger in appearance. You are not aware of any of this as it unfolds—the moon just looks big—but you can easily subvert the illusion by turning it upside down. If you are looking at the real thing or even a picture of the bloated horizon moon, invert the picture, or turn around and look at the moon between your legs. Because the horizon is no longer a reference point, you then see the moon as you would if it were overhead, and it seems to shrink.

By this reasoning we stand surrounded by perceived spheres of space of varying sizes, nested one inside the other. The nearest of these onion skins of space is where the (by now cold) cup of coffee sits. Determining the location of the cup in that space is essential for being able to pick it up, but so is an awareness of the location and position of the arms in relation to the cup. The brain relies on sensors in the limbs and the inner ear to measure both the pull of gravity and the direction of the limbs relative to the centre of your body. So if you are holding your arm straight out to the side (scarecrow-style) your brain knows where it is without looking because gravity is pulling the arm down and muscle sensors are confirming that the arm is at such-and-such an angle from the axis of your body.

The challenge then is to merge the map of your body (in whatever form it exists) with that part of the space map which includes the cup, the object of all this. The trouble is that there is an infinite number of combinations possible, depending on the circumstances of the moment. Look at it this way. You can sit in a chair watching morning television with your coffee cup beside you on your right. A sidelong glance at the coffee cup (with your head still firmly aimed at the TV set) establishes where it is, and you reach for it. But you can also move your head instead of your eyes, turning your head to the right until the eyes, staring straight out, light on the cup, and again you reach for it. In each case the reaching movement is exactly the same, but the image of the cup on the retinas of your eyes is different: in the first instance it falls near the edge, the second time right in the middle. Your head position was different as well, and something had to happen in your brain to allow for those differences and still ensure an accurate reach for the cup. It's all done with maps, but it's not done simply. Three examples from research in this area will give you an idea just how challenging this whole question is.

The beauty of frogs as neurological experimental animals is that you don't have to worry about stray thoughts from their cerebral hemispheres interfering with the pure task of homing in on an object. They don't really have cerebral hemispheres. Their brains are functional, as different from ours as the chip that runs your microwave is from the circuitry in a computer. But they will respond predictably to a target object in their personal space. Dr. Paul Grobstein of Bryn Mawr College in Pennsylvania has analyzed what goes on in the frog's brain when it is tempted by a target cup containing, not coffee, but a live "prey item." Obviously to be able to lunge and flick the tongue right onto the target, the frog has to be able to locate it precisely in three dimensions: direction, distance and height. Dr. Grobstein has been able to show that in the frog's brain, there isn't just one centre that handles all this information. At some places (and there aren't that many places in a frog's brain) the direction information is physically separate from the distance and height data. Damage to those nerve circuits carrying the direction information creates a strange scene: a juicy insect is placed off to the frog's right, and it reacts immediately by snapping at just the right height, with just the right lunge to cover the distance. But there's one slight problem—the frog aims straight ahead. No matter where the target lies, these brain-damaged frogs attack a point directly in front of them. They obviously see the target, because they respond as soon as it is put in place, and are aware of its height and distance, but have no idea in which direction it lies. Further experiments might reveal that height and distance are computed separately as well, but even the work done so far shows that locating and consuming prey is not the simple reflex action in frogs that it was once thought to be. It is tempting to try to imagine what this is like for the frog, and the simple answer might be that it is not like anything. The frog's tiny amphibian brain is not likely to be capable of reflection, and so will not stop to wonder why it didn't catch the insect. As far as anyone can tell, the frog's brain puts together the information that there is an insect out there and then directs the leg muscles to jump and the tongue to flick. If a mouthful of insect doesn't result, that's it. It slips into the waiting mode again, without puzzlement or regrets. After all, this is an animal whose eyes are programmed to perceive small, dark moving objects in preference to most other visual stimuli, and so is great at catching flies. On the other hand, a frog sitting in a cage with a pile of dead (and therefore unmoving) flies will starve to death. It can't see them.

So can we put ourselves in the frog's shoes? That would be very tricky: part of the problem is that we can never know why one of these brain-damaged frogs always aims straight ahead. Does the frog's brain "think" the insect is actually there and on that (incorrect) basis orders a jump in that direction? Or is there no directional information available to the frog's brain at all, with the result that it orders a straight-ahead jump because that's the "default" jump in the repertoire? If we were in the same position of experiencing the illusion that the target was in front, it would be a very unsettling experience to reach, grasp and come up with nothing but air. What would be just as unnerving (but is very difficult to imagine) would be knowing there is an object out there somewhere, being able to pinpoint its height and distance, but having absolutely no idea in which direction it lay. Would it be a hall-of-mirrors experience, with targets in every direction one looked, a kaleidoscope of the visual system? Or is it silly even to hazard a guess? There's an active debate among philosophers as to whether it is possible to know what it's like being a frog. If the frog's brain does not have self-awareness built in—as many suspect—then even the frog wouldn't be able to answer that question. While psychologists, animal behaviour experts and philosophers chew that one over, experiments with human subjects suggest there is some common ground between the two species.

At the University of Minnesota, Martha Flanders and John Soechting work with a different experimental animal: the undergraduate student. Although students are somewhat higher on the phylogenetic tree than frogs (or in a less speciesist view, on a distant branch), they have a lot in common, at least when it comes to finding a target. In these experiments students were shown a target briefly, then asked to point to where it had been after it was removed and the lights turned off. They consistently made errors, not of direction, but of distance: they sometimes underreached the target by as much as ten centimetres, even though at that point it had still been well within reach. An analysis of these errors showed that these students processed information about the direction of the target in an east-west sense separately from its distance and elevation—just like frogs. The idea that we still house a "reptilian" (or at least "amphibian") brain holds true, at least when it comes to reaching and grabbing.

Finally, psychologist Mel Goodale at the University of Western Ontario and his colleagues have found that there is a separation in the brain between knowing where an object like the cup of coffee is and being able to reach out and grasp it. Goodale has worked with two brain-damaged patients who are in effect mirror images of each other. One, known by her initials D.F., suffered damage to the visual areas of her brain as a result of carbon monoxide poisoning. If a set of oddly shaped asymetrical objects (like pebbles on a beach) is put in front of D.F. she is unable to tell by looking at them whether any two are the same. A second patient of Goodale's named R.V. who has damage to different areas of her brain had no trouble perceiving the differences or similarities among these objects. But there is a complete role reversal when the two patients are asked to pick these objects up.

Think about this the next time you are in a supermarket. If you are picking up irregular-shaped objects like potatoes, yams or pieces of fresh ginger in your fingers, you usually aim at two easy-to-grasp points, one on either side, and position your forefinger and thumb in a way that they will land on those points. It's another one of those things you do without thinking, but with great precision. R.V., the woman who has no trouble distinguishing between two objects like this, is nonetheless incapable of grasping them in the normal way. In experiments she often chooses inappropriate, unstable places to put her finger and thumb, and is only able to grasp an object firmly *after* she made contact. In other words, she can't make use of her ability to analyze these shapes visually when it comes to picking them up. D.F. is the reverse. This woman hasn't a clue when it comes to telling objects apart by looking at them—isn't even aware that any two are different—yet can pick them up just as you and I would, casually but unerringly choosing the right places to grasp.

These two patients support Goodale's contention that seeing the coffee cup on the table next to you initiates two separate mental processes. One deals with the nature of the cup itself, and would allow you to describe its dimensions, how it differs from other cups, and exactly where it is. The second stream of mental activity stimulated by the sight of the cup would set in motion the reach towards it. Both begin in the visual areas at the back of the brain where the simple image of the cup first takes shape, but they then proceed towards the front of the brain (and are elaborated) via different routes: reaching for the cup takes the high road to the top of the cerebral cortex, while knowing about the cup takes the low road along the temporal lobes. If one or the other pathway is damaged, it's possible to know about the cup but not be able to reach for it accurately (R.V.), or to reach and grasp it perfectly well without knowing, or at least being able to describe, the shape of that cup, or how it differs from others (D.F.). Mel Goodale is challenging the traditional view with this idea—formerly the two streams of information were designated as "what" and "where," whereas he is now suggesting they are "what and where" and "reach for it."

All this for a lousy cup of coffee. It is reminiscent of the comments of Tomaso Poggio, a pioneer in robot vision at the Massachusetts Institute of Technology, who has pointed out that while we are all too aware of the things our brains do least well, including recently learned (at least in the evolutionary sense) pursuits like mathematics and philosophy, we remain unaware of how truly powerful the brain is when it comes to simple (and ancient) skills like seeing objects such as cups. I would only add that there is nothing that is really simple.

8
The Little Man

Human brains capture our attention best when they do something spectacular, whether it is a Robin Williams monologue or a Stephen Hawking equation. Brilliance may sometimes even come from brains that are far from perfect. So-called savants (formerly known as "idiot savants"), retarded individuals who have one amazing skill, can name the day of the week of any date in history or play a difficult piece of music after one hearing. But these wondrous accomplishments are not what the brain does for a living. Daily brain work is concerned with the dreary but essential second-by-second maintenance and monitoring of the body. We seldom pay attention to this sort of brain activity; much of it never reaches consciousness at all. Unlike the thoughts of Hawking and Williams it has no voice. But as unspectacular—and unnoticed—as this brainwork is, it is programmed by some ingenious mental machinery.

One of the day-to-day tasks of the brain is to keep track of the position and orientation of the limbs, trunk and head so that the simple things in life like grasping a cup of coffee are possible. You don't have to will this to happen—your brain maintains an image of the body whether you are aware of it or not. You are free to pay attention to it if you like, but it won't wither away if you don't. Experiencing the body image is easy: with eyes closed you can envision exactly how all the parts of your body are arranged in space. Understanding how your brain creates such an image is, however, much more difficult, but it begins with a brain map of the entire surface of your body.

The map of the body in the brain is called the "sensory homunculus," meaning "little man." This unusual word has had a curious history in science. In the 1600s and 1700s it was commonly believed that each human sperm (some theories claimed it was the egg) carried within it a miniature person, which, upon fertilization, began to develop into a fetus. The theory was called preformationism; the little man, a homunculus. He was commonly depicted crouching inside the head of a sperm or egg with his arms wrapped around his knees, more cramped than those who rode the first space capsules. The theory was a powerful demonstration that what you believe can become what you see: in the late 1600s, the newly introduced microscope allowed several scientists to persuade themselves that they had glimpsed these miniature humans for the first time. Of course the homunculus in the sperm carried other homunculi in his sperm and so on and so on. By this reasoning the last homunculus had to be a member of the last generation on earth, and Adam and/or Eve must have carried the entire human race-to-be in their sperm or eggs.

The idea of a little person in every sperm was abandoned a long time ago, but it has reappeared in brain science as a little person representing the way our bodies are mapped in the brain. A strip of cerebral tissue running from side to side across the top of the brain receives sensations generated by touch receptors scattered all over the skin. On their way to the brain these sensations are sorted out so that those from the feet arrive on the homunculus next to those from the legs which in turn are adjacent to trunk sensations and so on. If each sensory neuron could produce a tiny image of the patch of skin it is "feeling," like a single pixel of a video picture, a pointillist representation of the surface of the body would spread across the surface of the brain. The homunculus is the electrical version of that. Touch your knee and the knee of the homunculus "lights up," but only in the sense of mass firings of brain cells, visible as spikes on a graph. In the brain, sensations are nothing but patterns of electricity. The homunculus is a collection of brain cells tuned to the body surface which create a portrait of the body, albeit with two significant differences from most portraits.

Because the right hemisphere receives information from the left side of the body and vice-versa, the body image inscribed on the surface of the brain is actually two half-bodies, each occupying one hemisphere with the toes tucked into the crevasse between the right and left hemispheres, each lolling on his/her back, head-down over the surface of the cortex. Tickle your toes and the sensations end up at the summit, the very top of this strip of brain; pull the seatbelt over your shoulder and the pressure of the belt is recorded at a point farther down the side. But a kiss on the lips! That arrives way down, just about even with the tops of your ears.

In addition, the body as represented on the brain is distorted: it is a cartoon human with an oversized head, huge lips and sausage fingers. The distortions reflect the fact that certain parts of our body, like the lips and fingertips, are more touch sensitive than others because they are more densely packed with touch receptors. Since all these receptors send their messages on to the brain, those parts endowed with large numbers of receptors require a larger amount of brain space than do other body parts. So in the brain, sensitivity to touch translates to area.

The result is the grotesque body of the sensory homunculus: its miniature trunk, arms and legs are dwarfed by an extended hand with a long pointing thumb. Beyond the tip of the thumb and detached from the rest of the body lies the homunculus's face, eyes and nose overwhelmed by the lips.

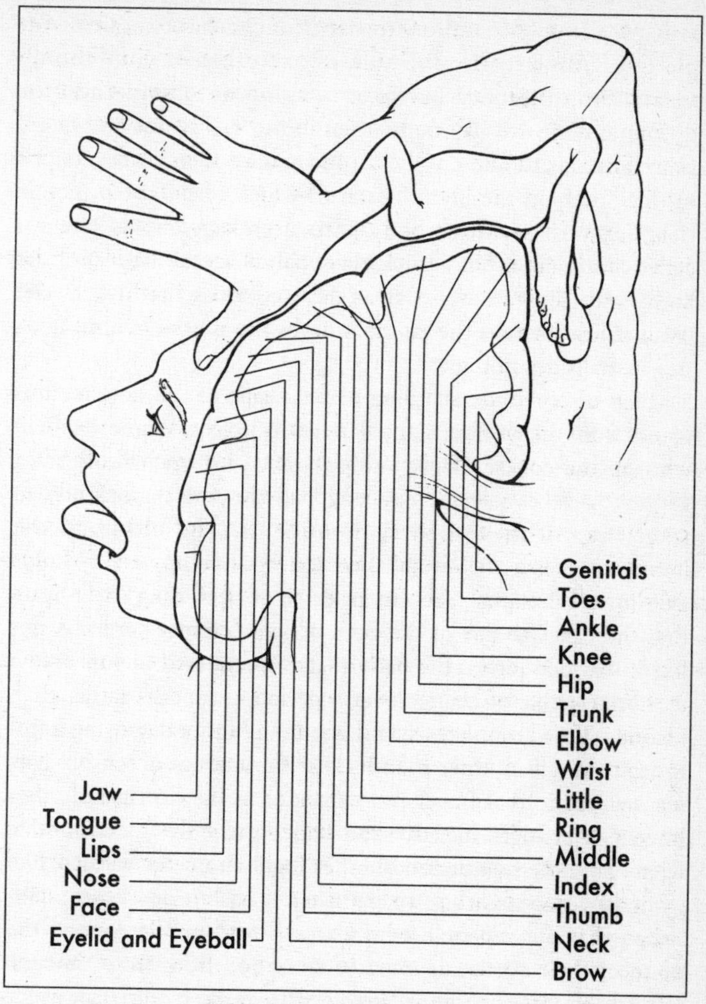

The Sensory Homunculus

The face is followed by the teeth, gums and tongue, all laid out one after the other.

The distortion of the homunculus makes perfect sense: more brainpower should be devoted to things of importance. It's rarely crucial for you to be able to differentiate between two touches a couple of centimetres apart on your back. But a couple of centimetres to one side or the other of your thumb means the difference between hanging onto something or dropping it. So we pay more attention to our hands and faces, especially our mouths. Other animals have their own versions of this "body-in-the-brain," each of which obeys the principle that the most sensitive body parts are exaggerated. The so-called "ratunculus" has a muzzle equal in size to the rest of the body, with separate brain areas devoted to each whisker. The batunculus portrays the animal's limbs flung back behind it, as they usually are in flight.

Much of our understanding of the shape of the homunculus comes from the work of Canadian neurosurgeon Wilder Penfield who, in the course of preparing patients for open-skull brain surgery to relieve severe epilepsy, had the rare opportunity to trace the outlines of their homunculi. And although the homunculus is a well-established concept that apparently holds true for all mammals, recent research is beginning to indicate that the exact layout of the body in the cerebral cortex is not fixed. In video terms, the body image is rendered in super-slow motion, capable of change over a period of months rather than seconds. If owl monkeys spend about an hour a day using a finger to twirl a disk, three months later the amount of sensory cortex devoted to that finger has expanded at the expense of other fingers. If changes like this can happen in only three months, what about the homunculi of other highly trained creatures, like gymnasts, trapeze artists or pianists? It has already been established that blind people who learn to read Braille expand the areas of their cortex devoted to sensation from those fingers. Would the intensive effort involved there be required in every instance? Maybe the homunculus of a one-fingered typist lies there on the surface of the brain crushed under one huge finger.

The homunculus illustrates an important principle of brain organization: it's not the nature of the incoming message, but where it plugs in that determines what you perceive. If you could eavesdrop on the sensory nerve messages arriving at the brain from your neck and your knee, you would be unable to tell them apart—each would be a train of electrical impulses, with nothing to identify the source. But the impulses from neck and knee, because of the wiring of the brain, arrive at different places on the body of the sensory homunculus, and it is that difference in location that allows you to know where you have just felt a touch. If the wrist of your homunculus were to be touched with an electrode you would feel a sensation, not on the surface of your brain, but in your wrist. That sensation can be enhanced by other mental processes, like expectation (feeling for the bottom stair in the dark) or dread (the brush of scaly skin in the cave); you can even "turn on" awareness of the homunculus. Take a moment to feel the clothing on your body. By doing so you become aware of many sensations that are present but unattended. Once they have been brought into the attentional spotlight it is very difficult to get them offstage.

Even the superficially simple business of becoming aware of a touch is rife with controversy. It seems simple: a fly lands on your wrist, the impulses travel from the touch receptors through the spinal cord to the wrist of the homunculus, you feel the fly's tiny feet and flick your wrist. It all happens in much less than a second. But that simple sequence of events has been called into question by a series of experiments performed by Benjamin Libet at the University of California at San Francisco, beginning in the 1960s. Libet had the unusual opportunity, as had Wilder Penfield before him, of working with patients whose brains were exposed during surgery, but who were nonetheless awake and alert. Libet was therefore able to ask them to compare two different sensations: one created by touching the patient's wrist with an electrode, the other by using the same electrode to stimulate the "wrist" of the sensory homunculus. If done carefully the patient should experience the same thing in both cases: a touch on the wrist, the only difference being that the direct stimulation of the wrist area on the homunculus shortcuts the normal route for such a sensation. Libet found that patients did indeed experience both as a touch on the wrist, but there was one big surprise. If he first stimulated the left side of the brain (which would feel to the patient like a touch on the right wrist), then touched the patient's actual left wrist, the patient reported feeling the sensations in the reverse order: a touch on the left wrist *followed by* a touch on the right wrist. This is completely bizarre: even if the electrodes made contact at exactly the same moment, direct stimulation of the brain should be detected first just because it takes time for nerve impulses from the wrist to get to the brain. But in these experiments the electrode was applied to the wrist about two-tenths of a second *after* it touched the brain, and even so patients detected, or became aware of, that touch on the wrist first.

These experiments are not easy to explain. Libet argues that there is a difference between the actual signalling events and our awareness of them, claiming that the signal from the wrist, even though it arrived later than the signal delivered directly to the brain, was "referred backwards in time." In other words, when we become conscious of it, it seems to have happened earlier than it literally did; it might be described as a case of our brains deceiving our minds. This interpretation has been greeted with great distaste; it seems to support a separation between the brain, the hard-working organ that perceives and records these events, and the mind, the place where we actually become aware of them. Most brain scientists today think that there is no difference between the two: the brain *is* the mind. So there is no shortage of critiques of Libet's work. Some point to the fact that the experiments have not been replicated by others; some argue that his experimental subjects might simply have experienced an illusion that one event preceded another, in the same way that we are all fooled—very convincingly—by visual illusions. Still other critics contend that it is actually impossible to fix the times we become aware of events because consciousness shuffles perceptions and thoughts like a Vegas blackjack dealer, and you never can be sure when you look at your hand which cards arrived first. But much of the criticism has been aimed at Libet's claim that the brain stamps a false time code on events before we become aware of them—the experiments themselves still cry out for some explanation, and at the very least they illustrate that feeling the world around us, or even the clothes we have on at the moment, isn't a simple act.

Even if Libet's experiments are taken out of the mix, it is still easy to demonstrate that feeling an object involves much more than simply detecting a signal arriving at the sensory homunculus. That is only the start: the brain then begins to interpret this information and may even attach meaning to a simple pattern of touch signals. Experiments in which people have lower-case letters traced on the skin with a wooden stick have revealed that a letter perceived as a "b" if it's traced on the front of the hand may be identified as a "d" if it's on the back. In fact this applies to the entire body: everything traced on the front is a mirror-image of exactly the same design traced on the back. But this effect can be mimicked even when the same body surface is touched—if you hold your palm facing away from you, the letter "b" is usually felt to be a "d," but if you then hold your palm facing you, that same tracing becomes a "b."

Now how can this be? A "b" on your palm is a "b" on your palm, and there can't be much doubt that the raw sensation of the letter is sent unaltered to its appropriate location on the sensory homunculus. But somewhere else, in a higher processing centre, that straightforward interpretation of touch is reconsidered, after factoring in information about the whereabouts—and orientation—of the particular piece of skin that was touched. Any delay caused by this is not apparent, but maybe that is because you didn't become aware of the touch until it had been fully examined by your brain.

An even more convincing demonstration comes from the condition called neglect, in which patients who have suffered brain damage (usually to the right side of the brain) lose awareness of the left side of space, a loss that often washes over the left side of the body. A neglect patient will sometimes fail to put make-up on the left side of her face, or even dress the left side of her body. Morris Moscovitch of the University of Toronto and colleague Marlene Behrmann have shown that the neglect patient's lack of awareness of the left side of the world applies even to simple touches on the hands, sensations you'd expect to travel directly to the sensory homunculus and make their presence known. Yet a touch on the right side of the wrist with the palm up—right where you take your pulse—will be felt by a neglect patient, while a touch a couple of centimetres away on the left side escapes notice. If the patient then flips her hand over, palm down, and the test is repeated, the touch on the right-hand side is again detected but the touch on the left is not. In other words, she would be aware of her baby finger if her hand was held palm down, but unaware of it when the hand was palm up. This makes the job of envisioning how neglect works a little tricky, because some parts of the body map are either noticed or not, depending on where they are, meaning that neglect applies to them sometimes but not others. How is it that an attentional spotlight could be directed to certain brain cells one moment but not the next?

The homunculus is one of the few examples of a map that is a neat, one-to-one (allowing for distortion) representation in the brain of some aspect of the outside world—in this case the surface of the skin. But a map is all it is, in the same sense that a map of Istanbul can enable you to find any street in the city but gives you no inkling of what it's like to be there. Your homunculus doesn't provide an experience—it is merely a guide to help you locate and identify touch signals.

This is not to imply that it is coarse, inaccurate or lacks versatility. In some instances your sense of touch has a remarkable ability to reach beyond its normal limits. Daniel Dennett, the author of *Consciousness Explained*, pointed out that you can explore the textures of the world around you very effectively using the tip of a pencil held in your fingers. If you drag the pencil lead lightly over the surfaces of objects like paper, wood, plastic or cotton, you usually have no trouble telling one from the other. This is amazing considering that what you're feeling is not the direct surface-to-skin receptor sensation you are used to, but the bumps and wiggles of the tip of the pencil. Yet these seem to translate effortlessly into feelings. You could go even further and hold the pencil between your teeth, in which case the vibrations of the pencil would be funnelled through the nerves of the teeth then to the brain to yield the same sensations. No one would argue that the sense of touch isn't very good at what it does. But in terms of creating an image of your body, it provides the groundwork and not much more.

The fact that the perception of letters traced on the hand may change depending on where the hand is facing suggests that one of the significant factors colouring the relatively simple perception of touch on your skin is the positioning of your arms and legs. That positional information flows into the brain from specialized receptors in the muscles all over your body. When you sit with your arms folded across your chest, these sensors do not tell the brain where your arms are—they simply record which muscles are contracting at that moment and which are not. It's up to the brain then to translate these measurements of muscular tension into arm and leg positions.

An important part of this positional information is derived from the tension of muscles resisting the pull of gravity, and when gravity goes, confusion sets in. Besides the well-known vertigo suffered by two out of every three astronauts, loss of a sense of where the limbs are is common too. Canadian Marc Garneau noticed when he was in the space shuttle that as long as his eyes were still closed when he woke up in the morning, he had no idea where he would find his arms and legs. Tests he participated in gave credence to that by showing that in the absence of gravity astronauts have great trouble estimating how bent their arms and legs are. In the 1960s one astronaut in the Gemini program, in which two American astronauts shared a space capsule, woke up in the morning and was startled to see luminous watch dials floating in front of his eyes. When he reached for them, they inched away, keeping pace exactly with his reach. The watches were on his own wrist. His arm had floated up to eye level in his sleep.

But even in normal gravity, the brain can be deceived easily about the location of various body parts. This is surprising given that it's hard to imagine a brain activity more fundamental or ancient than the maintenance of an accurate body image. One simple way to do this is to stand with arms outstretched in front, then raise one arm about 45 degrees and hold it there for half a minute. If you then close your eyes and try to return that arm to its original position alongside the stationary arm, you will find that you almost always leave it measurably higher. Conversely, if you lower one arm by the same amount, hold it there for a while, then try to raise it, you will again fall short. Over the ensuing few seconds your arm will involuntarily make several adjustments in angle which finally return it to the right place. (This little experiment works whether you're holding up your arms yourself or whether someone is supporting them for you. But remember if you try this that it will always work better with someone who doesn't know what the outcome is supposed to be.)

A different version of this is to close your eyes and turn your head to the side about as far as it will go without being uncomfortable and hold it (or have it held) there. After ten minutes or so you will have a strong feeling that your head has turned at least halfway back towards the front. These experiments make it clear that the brain's opinion of the whereabouts of the arms, legs and head isn't good enough: it needs to be calibrated by other information from the senses, especially vision. As soon as you open your eyes any mislocations are corrected instantly. As long as your eyes are open and you're in an earthlike gravitational field you will have a pretty good idea where your limbs or head are.

James Lackner and his colleagues at Brandeis University in Waltham, Massachusetts have devised laboratory experiments that extend the above exercises to the point where spectacular bodily illusions are created. They have found that vibrating a muscle or its tendon in the arm or leg will cause the muscle to contract, but if the movement that would normally be caused by that contraction is resisted, the person experiences the illusion that the limb is moving in the opposite direction. Normally information about the degree to which muscles are contracted is sent to the brain and used to construct an image of body position. In these experiments, whenever there was a contradiction between the position of a limb and body shape, limb position won out, with some very odd body shapes resulting.

Picture this: a volunteer is holding her nose while the biceps of that arm is vibrated, and she feels as if her forearm is being extended. But because it is apparent that her fingers are still in contact with her nose, and, as far as she can tell, her head is stationary, she is led to one of two conclusions: either her fingers or her nose (or both) have grown to keep pace with her extended arm. Some people in these experiments have become convinced that their noses have grown thirty centimetres long. If the opposite muscle in the arm, the triceps, is vibrated, the subjects experience the reverse sensation: that their noses are being pushed—painlessly—into their heads, or their fingers pass into their hands through their noses. In another experiment volunteers placed their hands on top of their heads, and when the triceps muscle was vibrated, they felt as if the hand was slowly pushing their head down into the torso.

It seems odd that the arms and legs should take precedence in establishing the body image to the degree that apparent movements of them lead to impossible distortions of the rest of the body. James Lackner suggests that the simple answer might be that we can check the positions and lengths of our arms and legs all the time by reaching for and touching things. But your nose doesn't have that opportunity, and so it needs contact with the mobile parts of the body—the arms and hands especially—to update its shape and size. This theory may not be right, but the idea that parts of our bodies like the chest, back and face are crying out to be touched just so the brain can reassure itself that they are the size and shape it remembers is somehow appealing.

9
Phantom Limbs

In Herman Melville's *Moby Dick*, as the ship's carpenter is busy creating a new artificial leg out of whalebone for Captain Ahab, he dares to ask the captain to verify a strange rumour he has heard:

> "... that a dismasted man never entirely loses the feeling of his old spar, but it will still be pricking him at times. May I humbly ask if it be really so, sir?"
>
> Ahab confirms that a lost limb may somehow linger on in the sensations felt by the amputee:
>
> "Look, put thy live leg here in the place where mine once was... Where thou feelest tingling life; there, exactly there, there to a hair, do I..."

Moby Dick was first published in 1851; little more than a decade later a second reference to this unusual subject was made in a piece of fiction published in *The Atlantic Monthly* called "The Case of George Dedlow." Dedlow was supposed to have been a surgeon who had served with the 79th Indiana volunteers in the Civil War and had suffered so many wounds and subsequent complications that he had become a quadruple amputee. The story relates Dedlow's detailed observations of his and other amputees' continuing experience of their missing limbs: how itching or painful sensations seemed to come from the missing limb with easily the same intensity and quality as they had before the amputation; how an amputee could feel the exact position of an absent arm or leg, even to the degree the fingers or toes were flexed; how sometimes a foot or hand, although vividly present, seemed attached directly to the next joint up, the knee or elbow, or even to the stump itself. In one bizarre scene Dedlow attends a seance in which the medium transmits a message from the spirit world: "United States Army Medical Museum, Nos. 3486, 3487." Dedlow realizes with a gasp that these specimens are his legs.

The author of *The Atlantic Monthly* story was unidentified, but it is now known to have been Dr. S. Weir Mitchell, a Philadelphia doctor who worked in a hospital where most of the patients were Civil War amputees. In the years following the Dedlow story he published a second account (in 1871) in a general audience magazine called *Lippincott's Magazine of Popular Literature and Science* in which he referred to the sensations of missing arms and legs as "phantom limbs," the term still used today, and it wasn't until 1872 that Dr. Mitchell finally described phantom limbs for his fellow doctors in a neurological text. Nobody seems to know why he approached the subject so delicately, writing first anonymously, then in a popular magazine (while still not admitting authorship of the *Atlantic* article) and only then for his colleagues. Some have suggested that the subject was too indelicate (while that would account for anonymity, it doesn't explain why he wouldn't have written about phantom limbs in a medical journal); others suspect that Mitchell had written the story for his own amusement but that it had circulated a little too widely, into the hands of the *Atlantic* editors, who decided to print it and paid Mitchell eighty dollars. By this account the story in *Lippincott's*

Magazine was intended to correct some mistaken impressions the public held from reading the *Atlantic* article. Regardless, it was an unusual way to bring a medical subject to light; writing about a medical phenomenon for the lay public first would be unheard of today.*

McGill University psychologist Ronald Melzack is one of the world's foremost authorities on the causes and treatment of pain, and as an inevitable consequence of learning about pain, he has explored the strange world of the phantom limb. Almost all amputees continue to experience their missing limb, and many of those are unlucky enough to feel pain, sometimes unbearable pain, coming from the phantom. Far from being a simple continuation of nerve activity in the stump of the amputated limb, phantoms appear to be produced in the brain.

One of Melzack's recent case histories makes that clear. He and his colleagues are detailing the experiences of a girl, K.G., who had had her leg amputated below the knee when she was just six years old. Shortly after, she was fitted with a prosthetic lower leg. She has told the researchers that soon after the amputation, she began to feel the presence of a phantom foot in roughly the same place where her amputated foot had been. This phantom, although not complete (the sides of the foot and the toes are there, but there is no top or bottom) is capable of some movement—she can wiggle its toes—and sensation—it feels hot if she holds it near a stove or fireplace. But the most extraordinary feature of this phantom foot is that it is not alone.

She has a second phantom, a set of miniature toes seemingly attached to the bottom of the stump, and as small as the minia-

* We live in times when medical scientists won't even talk about what they've just submitted to a medical journal for fear that the editors will somehow get wind of their indiscretion and refuse to publish their work in the future. Once the journal embargo is lifted, however, everyone gets the story: *The New England Journal of Medicine,* one of the highest profile medical journals in the world, changed the time of release of their top stories of the week a few years ago so as to allow television journalists to get the stories on the US evening news programs.

ture toes are, they are also the most irritating, constantly itching and wiggling. The wiggling at night-time makes it hard for K.G., who's now in her twenties, to go to sleep. There is even a third phantom, a leg and a foot which come into existence to fill her artificial leg when she puts it on. This third phantom is the least vivid, but its foot generates a variety of sensations: it can be tickled, it sometimes feels sweaty, and most unpleasant, seems fixed in one position. This is an intriguing feature, because that foot prior to amputation had been locked in place too, but K.G. has no memory of that. As far as she can remember, she had been able to move that foot normally. So the fact that this immobile phantom resembles the real foot that preceded it cannot be attributed to any sort of conscious memory that K.G. has. There is a memory, but in a sense it's her brain's memory, not hers.

This case has all the features that make phantom limbs a hot topic among brain scientists. First there's the very existence of a phantom, a part of the body that is imaginary—whatever that means in this context—but which may experience all the sensations a flesh-and-blood one would; a nonexistent limb that feels absolutely real. Wondrous to the observer, but rarely to the owner. K.G. is bothered by both her second and third phantoms, but there are many amputees who suffer much more than she does.

The majority of people who possess a phantom also suffer phantom pain, and the descriptions make clear just how awful this can be. It is described as knifelike, burning, crushing, twisting and grinding. "It feels as if someone is trying to pull your leg off"; "Like an electric shock"; "As if someone is sawing it off." Patients have felt as if their thumb was being pushed through the palm of the hand, or that the fingernails were being lifted from their beds. From descriptions like that it is clear that these are not merely mimicking some previously experienced sensations—these are wholly new and horribly vivid pains coming from a nonexistent limb or hand. Amputees have committed suicide as a final escape from constant, searing pain. And sadly most attempts to alleviate the pain are unsuccessful. Implanting electrodes, stroking or applying pressure to the stump, hypnosis and/or anaesthetics—nothing is very effective. Even cutting out the piece of the brain where these pain sensations should be arriving has been tried, mostly without success.

Even in the absence of pain, phantom limbs are extraordinarily vivid. The story is told of a patient who had lost his right arm below the shoulder, yet still felt as if his right hand were there. Just as he reached for a cup of coffee with his phantom hand, the investigator suddenly pulled the cup away and the patient cried out in pain. To him it had felt as though the cup was being torn out of his (nonexistent) fingers. Pain may be the most significant sensation accompanying a phantom, it is far from being the only one as the case of K.G. illustrates. Phantoms can itch, feel wet, prickly, warm or cold. A ring once worn can still be felt on a phantom finger, a watch on a phantom wrist. In addition to "remembering" details like jewellery that was once worn on the amputated limb, the sensation of a phantom may continually be updated. If the possessor of a phantom becomes chilled, he may feel that his phantom has goosebumps. One patient with Parkinson's disease began to experience the characteristic stiffness, awkwardness and tingling in his phantom left arm as the disease progressed.

Usually the phantom behaves normally—an arm hangs by the side when the person is standing, but swings back and forth when they begin to walk. When an amputee sits in a chair, the phantom leg bends at the phantom knee. However, this isn't always the case: in one famous instance a man thought his phantom arm stuck out from his side at right angles, and so he always turned sideways to walk through doorways. There couldn't be a more vivid demonstration of the reality of a phantom limb; after all, he only had to look to see that he no longer even had an arm.

Phantoms of the arms and legs are by far the most common, but there are reports of a phantom nose, phantom breasts after mastectomy, and a phantom penis. On the other hand, there appears never to have been a report of a phantom ear.*

The case of K.G. is important because she was only six years old when her lower leg was amputated. There have been claims that children who are operated on at this early age will not form phantoms, either because they didn't have the limb long enough to make it part of their body image, or because their young and still developing brain will quickly reassign the sensory area vacated by the limb to the reception of completely different sensory information.

Imagine the area in the sensory homunculus of K.G.'s brain that was devoted to the right leg. What would happen to that area after the amputation? In the short run, it would fall silent because there would be no more sensations coming to it. But there is a growing body of evidence suggesting that this situation would be temporary. The brain can't afford to have a patch of cells in the cerebral cortex with nothing to do. Experiments over the last ten years or so have established that brains in general (including the human brain) are much more flexible and less hard-wired than they were once thought to be, and that the sudden removal from one part of the brain of all sensation from an amputated leg would be a little like exposing a patch of garden soil by tearing the flowers out—it wouldn't be long before other plants had taken root. This is an inexact analogy, because the brain cells in the leg area of the homunculus would remain unchanged; the difference would be that they would be receiving signals from a different part of the body, by virtue of having established connections with a replacement set of sensory neurons.

In the case of a missing arm it might be expected that either nerve cells reporting from the face or the shoulder might

* Presumably the surgical reattachment of a severed body part would occur too soon after amputation to allow the formation of a phantom, but this does raise an interesting question. If there were a phantom, then the part was reattached, would it replace, extend or co-exist with the phantom? And yes, I was prompted by the Bobbitt case to ponder this question.

invade the now uncommitted brain space, because those are the areas that lie immediately to one side or the other of the arm on the sensory homunculus. In fact, there have been cases where that is exactly what appears to have happened; in some patients it is possible to create a sensation of a phantom arm by touching either the area of a stump near the shoulder, or, strangely enough, the face. It's even been possible to map a phantom arm directly onto the face piece by piece: the wrist under the chin, the elbow lying along the jaw and so on. Such replicas of phantoms seem to be proof that the territory in the brain formerly held by the amputated limb is colonized without delay. These experiments are a puzzle though, because the biological rationale for taking over the vacant space in the brain should be to maintain the maximum possible sensory reception in the brain. After the amputation of an arm it would make biological sense to recruit the area formerly devoted to the arm for facial sensations, and at the same time dispense with all sensations coming from nerves that used to monitor that arm. It makes no such sense to have facial sensations with a phantom arm superimposed on them. The very existence, let alone the persistence, of phantoms is proof that the picture is not exactly straightforward. In this case cited above, the arm persists as a kind of ghostly image shadowing the face. The explanation could be that we are seeing the transition between the image of the arm and its replacement, the face. But then what about cases like K.G.'s, where phantoms persist in a vivid and stable form for years (in her case fourteen), long after rewiring of the homunculus should have taken place?

Anomalies like this have persuaded Ronald Melzack that the image we have of our own bodies is only partly the creation of our senses. They are obviously important: the feel of your body at every moment enables you to close your eyes and describe the position of your trunk and arms very precisely. But in those rare instances where there is a shortage of sensory information, a body image persists regardless of any rewiring that occurs, and Melzack suggests this is because there is a permanent representation of it in the brain. He envisions three parts of the brain co-operating to maintain what he calls a "neurosignature" of the body. Besides the sensory homunculus, Melzack believes that the limbic system, the part of the brain responsible for emotions, and also the parietal lobe are involved in maintaining the complete image of the body.

Melzack is convinced the limbic system must be involved because paraplegics whose spinal cords have been severed still describe sensations from the phantom lower parts of their bodies with the same emotional terms ("pleasurable," "exhausting") as they used before their accident. The parietal lobe plays a key role in defining in spatial terms what is and isn't us, and damage to this part of the brain may result in inattention towards, or bizarre beliefs about, a part of the body (see Chapter 10). In Melzack's view the combination of touch sensations, emotion and a three-dimensional impression of body-in-space creates a body image so powerful that it can harbour phantoms.

Even though amputation might result in the complete elimination of certain sensations from the flow of information into the "neurosignature," the brain image is fully capable of carrying on. In Melzack's view, activity in any part of the network creating the "neurosignature" would tend to preserve the former body image. A phantom limb, by this explanation, is simply a fabrication of the brain. It resembles the real limb because for years it represented the real limb. In that sense it is more like a neural portrait. As long as the model (the limb) is present, the portrait is stunningly accurate. Once the model leaves, the portrait slowly but surely begins to change. So there are phantom limbs that start out at full dimensions, but gradually telescope, producing a hand attached to a shoulder, or K.G.'s second phantom, the set of miniature toes at the end of her stump. But the neural portrait is at least partly independent of whatever is going on in the sensory homunculus, and need not disappear abruptly either as a direct result of the loss of a limb or the more gradual process of rewiring. It might even be said that the mind's portrait of the body is in some sense *completely* independent of incoming sensory information—how else to explain the fact that some people born with a missing limb may still experience a vivid phantom of that limb?

Strange as they sound, phantoms are not nearly as far removed from everyday experience as you might think. Ronald Melzack has shown that anyone can experience a phantom if the circulation to an arm or leg is cut off and the limb falls asleep. He managed to do this in the lab by using a blood pressure cuff to cut off circulation to a volunteer's arm. This person was seated at a table, arm under the table and hidden by a black cloth, and every once in a while was asked to mark the exact position of the arm at that moment. Melzack found that after sensation in the arm had disappeared, these people began to experience a moving phantom: the arm—which had been extended—was gradually bending at the elbow, the hand getting closer to the chest minute by minute. They weren't experiencing the feeling of movement so much as the sense that one minute the arm was over there, now here.

It took about half an hour for the arm to feel as if it had bent a significant amount. Some of the volunteers in the study apparently were astonished at the difference between where their arm actually was and where it felt as though it should be. The moment the cloth was pulled back and they could see their arm, the phantom vanished.

When enough time had passed, and the phantom had completely separated from the real arm, these people were asked to point to a target. The target was positioned to be exactly halfway between the phantom and the real arm. So typically the phantom arm was pointing to the left, the real arm to the right with the target between them. Most of these people looked at the target and directed their phantom arm to move to the right to point at it. Of course the real arm did the moving, shifting even further to the right and away from the target.

Most owners of phantom limbs have suffered no brain damage, and are completely aware that the sensations they experience are unreal, even though they may absent-mindedly still behave as if the limb were still there. In that, they are completely different from patients who, as a result of a severe stroke, are incapable of believing that a paralyzed arm or leg is actually their own. (More about them in Chapter 10). But there is a parallel between the two situations in that many patients who reject a paralyzed limb insist that the "real" arm or leg is still there, although usually hidden from view. They can feel it, they claim to be able to move it, even to *have* moved it on command. Even in the midst of their delusions, there is some sort of body image that maintains a feeling of how the now-paralyzed arm once felt.

How phantom limbs are created is not as important as the fact that they exist. They serve as another reminder that the brain is perfectly capable of inventing an experience that is in every way as vivid and convincing as the real thing. As long as we have to rely on the brain, we are simply not capable of knowing whether what we think we are seeing, smelling, hearing or remembering is real, close to reality or some hastily made-up re-creation of reality. But what else can we rely on?

10
"I'M QUITE SURE IT'S NOT MY ARM"

Dr. Sandra Black, a neurologist at Sunnybrook Hospital in Toronto, sees hundreds of stroke patients every year. One out of every forty or fifty such patients will display the bizarre symptoms of "asomatagnosia": they deny ownership of their own arms and legs, and make up exotic stories to account for the presence of these "alien" limbs. It seems totally bizarre to onlookers, but from the patient's (and the injured brain's) point of view, these fables might simply be an attempt to make sense of the information available to them.

In one such case, a woman in her sixties, a nurse, was admitted to Sunnybrook after complaining of a headache and developing sudden left-sided paralysis. She had suffered a large haemorrhage on the right side of her brain. Four weeks later she was alert with no apparent loss of intelligence, but her left side was still completely paralyzed. She was, according to the standard tests, also suffering from neglect, but while she didn't seem to perceive the left side of space when bisecting a line or crossing out lines scattered all over a page, she was aware to an abnormal degree of her paralyzed left side. She consistently denied that her limp left arm was actually hers:

> "Do you have any idea whose arm this may be?"
> "Yes I've looked at the tape (the hospital ID bracelet) on her wrist; she has a similar name to mine. It makes people convinced that it is mine."
> "What is the other name?"
> "H-Y-N-E-K" [The name has been changed to protect this patient's anonymity.]
> "What is your name?"
> "H-Y-N-E-*S* (emphasizing the "S"). They both have the same first name." (Apparently she misread the bracelet).
> "Who is Mrs. Hynek?"
> "Before I was on the ward, there was apparently someone called Mrs. Hynek in my room." (There wasn't.)
> "How do you imagine her arm got onto your body?"
> "I don't feel it's really onto my body. It lies on top. It's very heavy, very annoying."
> "Did her arm stay behind somehow after she left the hospital?"
> "It's the only thing I can suspect. I can't figure out how this happened."

Mrs. Hynes, being a nurse, was even willing to speculate that this arm, which had originally belonged to Mrs. "Hynek," survived because of the blood that was left in it, and indeed she was certain that there was none of her (Mrs. Hynes's) blood in it even now. She admitted it seemed weird, but added, "I'm not a person who believes in the supernatural. It doesn't make sense, but I'm quite sure it's not my arm."

With the help of the medical staff she examined this foreign arm and hand in detail. What she found supported her belief that it wasn't her arm. The fingers were thicker and the shape of the fingernails and the pattern of their ridges did not, in her eyes, match those on her right hand. The wedding band, although coincidentally a "similar style" to hers, was clearly different. She hated to think that her wedding band might have found its way to someone else's hand.

In Mrs. Hynes's mind, the fictitious Mrs. Hynek got the worse of the deal, because she left behind her arm and didn't get one in return. Mrs. Hynes's arm was apparently still attached to her, hidden under the arm that had been left behind:

> "When you say your 'own arm,' what do you mean?"
> "I mean in my mind, the true left."
> "Where is that 'true' arm now?"
> "Someplace between my shoulder and this extra arm. It often feels as if it's under it or just slightly over it."
> "Do you ever actually see it?"
> "No, just the feeling."

Through all this, Mrs. Hynes seemed indifferent to the indescribably strange events she was certain had happened. She didn't like the extra arm, finding it heavy and annoying as it lay on top of her, but she seemed just as annoyed, if not more so, that the medical staff seemed not to believe her story and kept trying to convince her it was her arm. Gradually, over the next two months, she came to accept that it was her arm, and that it was paralyzed.

The strangest aspect of this story is the complexity of the delusion. It begins with denial that the paralyzed arm belongs to her and is then elaborated into a story that the arm actually belongs to someone else, a delusion that is fully supported by (false) evidence that the left hand and wedding band differ from her "own." At that, Mrs. Hynes stopped short of doing some of the truly baffling things that other patients have done, such as becoming so annoyed with the foreign limb that they actually attack it, kicking the left leg with the right, or trying to throw the left arm out of the bed. These patients sometimes have to be restrained to prevent injuries to themselves. Sometimes they personify the paralyzed limb, often giving it a demeaning nickname: one of Sandra Black's patients called her own left arm, "Lefty," and criticized it for being a helpless baby that couldn't do anything for itself.

Some of the name-giving is accompanied by a kind of cheerful ignorance of the seriousness of the condition ("Lucky Joe," "Dolly Gray" and "Lazybones"), but others express the hostility and anger sometimes felt ("the Dummy," "a piece of dead meat," "a stick of wood"). There is even the rare lighter moment: in one case described in 1913 a male patient had concluded that his left side actually belonged to an erotic woman who was lying in the bed beside him. He used to make witty remarks about her, and occasionally would caress his limp left arm. Two weeks later the delusion had vanished, and the patient could only vaguely describe it from memory. A physician with strong musical likes and dislikes called his good right arm Chopin, and his paralyzed left Schumann, because it "wasn't much good." Another called his arm "The Communist," because it refused to work.

But wit is not the dominant theme with these patients. The woman who called her arm "The Dummy" referred to herself as the "broken doll." More often the paralyzed arm—the arm they no longer acknowledge as their own—is seen as disfigured, shapeless and very often snakelike; snakes appear again and again in these descriptions.

It is not uncommon for patients like Mrs. Hynes to suffer simultaneously from neglect of the left side of space and obsession with her own left side, specifically her arm. This seems paradoxical: why wouldn't she have neglected her own paralyzed left side? About all that neuropsychologists can say so far is that while the neglect and asomatagnosia often co-exist, it is possible to have one without the other, suggesting that spatial maps and body maps are different and separate and therefore can be selectively damaged by a stroke.

When all the recorded cases are put together, a spectrum of attitudes appears, ranging from an apparent total lack of concern and interest (sometimes including assertions that the arm or leg actually is moving) to indignant, even angry denial of paralysis and possession of the limb, to attempts to get rid of it. The patient's belief in what seems totally bizarre and senseless to us is almost unshakeable: in one case described early this century a distinguished scientist was so convinced of his normalcy that he doubted the sanity of his niece who was insisting (correctly) that he was paralyzed.

The delusions, when they exist, are detailed and complete. Most of these people, including Mrs. Hynes of the story above, argue that there are two arms. One is the limp one belonging to someone else, the other is their own arm, usually hidden from view under the "alien" one. They are perfectly capable of reporting where the "real" arm is and will even hallucinate feelings coming from it, even though it is of course paralyzed and insensate. They not only invent sensations where there aren't any—sensory information that is real is made to fit the story that is being created, not the other way around. Patients like Mrs. Hynes can examine details of the hand, fingers and even jewellery and be convinced that they aren't theirs. One male patient of Sandra Black's knew the arm wasn't his because it smelled like an animal. I know of no cases where a patient who rejects a limb has been persuaded either by visible evidence or logic that the limb really does belong to them. What seems necessary is for time to allow the brain to recover to the degree that these delusions disappear. In the meantime pointing out logical inconsistencies usually cannot shake the patient's view of the situation:

> Doctor: (takes the patient's paralyzed left hand and moves it over to the patient's right side) Whose hand is this?
> Patient: Your hand.
> D: (places the patient's left hand between his own hands) Whose hands are these?
> P: Your hands.
> D: How many of them?
> P: Three.

D: Ever seen a man with *three* hands?
P: A hand is an extremity of an arm. Since you have three arms, it follows that you must have three hands.
D: (holds his own hand up in the air) Put your left hand against mine.
P: Here you are. (Of course his paralyzed left arm remains motionless.)
D: But I don't see it and you don't either!
P: (after a long pause) You see, doctor, the fact that the hand didn't move might mean I don't want to raise it. My words may astonish you, but there are bizarre phenomena. My not moving my hand might be due to the fact that if I keep from performing this movement I might be in a position to make movements which would otherwise be impossible. I am well aware of the fact that this seems illogical and uncanny...

It's a shock to hear these stories coming from people who are, by all other measures, perfectly sane. They see nothing peculiar about suggesting that a paralyzed leg actually moves, or that their left arm actually belongs to someone else. The difficulty in understanding what is going on in the brain is that there are several different problems involved, ranging from the patient's lack of awareness of and/or concern for his own paralysis to the full-blown delusion of an extra limb. Most of these can occur on their own: some patients are merely unaware that an arm or leg is paralyzed, while some even admit to a paralyzed arm but deny a paralyzed leg. Others come as close as possible to admitting while still denying: "It is curious, it is as if I had been paralyzed." At the other end of the spectrum are stories like that of Mrs. Hynes.

To explain the variety of combinations of symptoms, Daniel Schacter of Harvard University envisions a sort of awareness network in the brain that is an interconnected set of modules not unlike a stereo system. It has an executive system—the equivalent of the tuner—that organizes and monitors all the other modules—CD player, tape, television. Damage can occur either within an individual module or in the central executive itself, with the result that a range of severity of symptoms can occur, as described above. In the stereo the symptoms can vary as widely, from no sound at all to fuzziness in one speaker; the causes can reside within the modules (dust in the CD player) or in their connections to the executive (the tape deck isn't plugged into the tuner). Presumably the network in the brain has important modules in places where damage often results in lack of awareness and/or denial, like the parietal lobes and the frontal lobes.

But there is much more to this story. For one thing there are the elaborate delusions and the stories made up to accompany them, both of which extend far beyond a mere lack of awareness. Second, the majority of these cases result from damage to the right side of the brain—what that means no one knows. Delusions and right hemisphere damage both play a part in stories like that of Mrs. Hynes, and the study of each contributes a piece to the puzzle.

The Harvard psychologist Brendan Maher has developed a somewhat counter-intuitive theory which begins by saying that there really is nothing abnormal about delusional thinking. He is convinced that delusions are theories just like scientific theories, invoked to make sense of something that is puzzling. Once the scientist, or the patient, arrives at a satisfactory explanation for the puzzle, she is extremely reluctant to entertain contradictions. (Remember Mrs. Hynes's impatience with those who doubted her.) Just as in science, data that supports a delusion is embraced, data that doesn't is ignored or rejected. (Even a wedding band can be seen to be slightly different from one's own.) What sets misguided thinking apart from the legitimate theory is the puzzle that led to it in the first place. Theories are seen by others to be nonsensical when they are based on information only the patient seems to have, or on information that everyone shares but only the patient finds puzzling. Mrs. Hynes had suffered a serious brain injury as a result of her stroke and so was unable to understand the lack of sensation from her paralyzed arm with the same ease as her nurses. She was therefore the only one who needed to come up with a theory to explain that derangement of feeling, and she was the only one who thought that someone else's arm was attached to her. But given that starting point, her theory is carefully thought out and logical. The important fact to note is that the full-blown delusion only needs a single wrong step to set it off. "That is not my arm" leads inexorably to the whole bizarre story of where the arm came from, how it survived and why it happened. It may even be that the story is so convincing to its creator that it alters how she perceives details in her wedding ring and her fingernails. Rather than the brain injury altering her visual perception so that the arm doesn't look like hers, the theory that it is not her arm might be so compelling that it dictates how and what she sees. And it doesn't take anything as serious as a stroke to set off delusions.

In one Danish experiment more than thirty years ago, subjects were asked to sit down in front of a large glass-fronted box, put on a glove and put the gloved hand through a hole in the box. At a signal they were then to pick up a pencil inside the box and trace a straight line on a piece of paper. They were able to watch all of this through the glass front of the box. Unbeknownst to them, however, an assistant was standing unseen directly behind the box, an identically gloved hand holding an identical pencil over an identical straight line. A movable mirror made it possible to switch what the subject saw from his own hand to the hand of the assistant, without his being aware that a change had been made. After a few trials with subjects watching their own hands, the mirror flipped, and the assistant, whose hand was now being watched by the subject as if it were his own, imitated the movements of the subject's hand until about halfway along the line, then started to deviate from the line. Imagine you're the subject: you feel as if you are guiding your hand straight but the hand you are looking at—which you justifiably assume is yours—is starting to slide to the left or right. As soon as the hand reached the end of the line, the lights in the box went out and the subjects were asked for their reactions. Many of them claimed to have *felt* their hand moving in the wrong direction, even though the trickery was entirely visual. In one version of the study reported by Brendan Maher, subjects also offered explanations like "I was hypnotized," "My hand developed automatic motion" and "There were electrodes on my hand but I could not see any; they were there, but I was deceived about them." That latter comment in particular sounds no different from the paranoid delusions of schizophrenics. If it is that easy to prompt delusional explanations, Mrs. Hynes's story about the arm of another person seems much less out of the ordinary: she was faced with a puzzling situation and created a sensible theory to explain it. However, given that even she realized that the story was bizarre (she said, "It doesn't make sense"), why did she stick with it?

This might be where the injury to the right hemisphere comes in. A number of experiments have established that a damaged right hemisphere is associated with an inability to see through illogical or even nonsensical statements. In studies where people listened to a brief story being told and were then asked to retell it in their own words, people with damaged right hemispheres embellished the stories much more than people with intact right hemispheres, occasionally adding completely irrelevant sidebars to the main story. Sometimes they claimed—maybe justifiably—that these improved the stories. One patient, when retelling a story about a grocer who had been robbed, changed the stolen money to groceries, and added, "Tripped on a bad sidewalk, fell down and broke all the eggs." It's tempting to suspect that compulsive liars may have some right-hemisphere problems, but it is very risky to extrapolate from these patients, who have serious brain damage, to people who lie a lot but are otherwise mentally intact. There just isn't enough known about what's going on in a damaged brain to make easy correlations.

But it wasn't just that these patients invented parts of their stories. When faced with the occasional incongruous sentence in the story (deliberately inserted by the experimenters) they went to great lengths to justify that sentence. So in the story of a farmer who gives his hired hand a raise after finding him asleep in a haystack, right-hemisphere-damaged patients offered the following explanations: "The cost of living is up..."; " Maybe he thought he wasn't paying him enough..."; "...to encourage him to work a little harder"; or even more elaborate, "The guy quits—no discussion or anything—the farmer meets him and raises his wages. 'Say hey, Joe—if you're gonna quit, I'm raising your wages. It won't cost me a cent, you won't be here!'"

It can't be said with certainty that both the embellishments to the stories and the farfetched explanations for incongruous sentences are products solely of the left hemisphere. It's tempting to think that damage to the right hemisphere unmasks the workings of the left, but it could be that these storytelling inconsistencies are what you get from a damaged right hemisphere. In any case, the tendency to invent an elaborate story, together with the ability to justify any statement, however odd, provides fertile ground for the creation of elaborate delusions. In that light, the delusions that may accompany paralysis and neglect of the left side and their unquestioning acceptance by the deluded patient may have their roots in these storytelling abnormalities.

Further experiments of a similar kind have established that these patients are unusually rigid when it comes to making sense of something they've heard. Asked to explain the meaning of the two sentences: "Sally brought a pen and paper with her to meet the famous movie star. The article would include comments on nuclear power by well-known people." Patients with damaged right hemispheres often stick with their interpretation based on the first sentence alone, and argue that Sally was going to get the movie star's autograph. They understand each individual thought, but they can't alter that understanding based on the larger context of the story.

If you think back to the tortuous logic used by patients trying to explain the existence of someone else's arm attached to their body, you can see hints of this same phenomenon. This may also explain why, when these patients recover to the point where they finally acknowledge ownership of the paralyzed limb, as many do, they seem incapable of remembering just what it was like to believe the unbelievable. They can remember having odd beliefs, but not the feeling of what it was like to have those beliefs. If their inability to follow the sense of a story was a transient problem, a transcript of their own remarks from that time would likely make no sense to them.

There have been suggestions that the explanation for these delusions is denial, the idea being that the psychological shock of having a limb or even a whole side of the body paralyzed is so great that these people simply can't admit to themselves or anyone else that it has happened. Of course this denial would be an unconscious process that would then be acted out by the patient. Indeed this same argument has been made to explain neglect—that complete inattention to a paralyzed limb could only result from the mind shutting it out. But there are some indications that this can't be the whole explanation: for one thing, the insistence that an arm belongs to someone else usually appears immediately after the brain injury, when the patient is still confused, and then fades as the patient becomes aware of the enormity of the situation. If denial were the root of the problem it wouldn't be expected to fade as awareness grew—exactly the opposite. It's also true that some patients deny the existence of one problem, like a paralyzed arm, while at the same time admitting difficulties with speech.

A variety of explanations are put forth by experts, and as is the case with many of these stroke-related problems, there is little or no input from the patients themselves. As we've seen, they are either uninterested or seem not to notice that there is anything amiss; by the time they've recovered sufficiently to take notice, their delusions have faded. They seem to have belonged to a different person, which, if the brain really *is* the individual, is not far from the truth.

11
Out of the Body

How far can the brain's image of the body be distorted? You can bend it momentarily by simple tricks such as those mentioned in Chapter 8, like trying to lower an upraised arm to horizontal with eyes closed. Most people have great difficulty sensing where the arm is—the brain's image of the arm has lost its usual precision. Such changes are trivial compared to those which can result from brain damage. Stroke and tumour patients may have their body image warped so much that they will be unaware that a limb belongs to them, and may even actively deny it. There is a haunting report from the beginning of the century of a woman who had lost track of her entire body so completely that she asked to be taken outside and thrown into the snow—she craved sensation, however unpleasant, that much. But there is a much more commonly reported phenomenon that, if real, would represent the ultimate in loss of body image: the out-of-body experience, the sensation of leaving the body behind and travelling wherever one desires.

When I use the cautious "if real" I am questioning the reality of the most spectacular forms of OBEs, those in which people report having travelled to distant places and witnesses are said to corroborate the claim. One of the best known of these is supposed to have happened in 1774 in Naples, when Alphonsus Liguori, founder of the Redemptorist order of the Roman Catholic church, fainted while preparing to celebrate mass. When he revived, he told those around him that he had been at the deathbed of Pope Clement XIV in the Vatican, a four-day journey away. News of the Pope's death arrived a few days later together with reports that St. Alphonsus had actually been seen at the Pope's bedside and had conversed with others there. Readers with any sort of skeptical bent will surely be tempted to throw that story out on the compost heap of other strange-but-true accounts, but bear in mind that stories like that are rare and represent only one extreme of the OBE spectrum. At the other are the accounts that I think must be believed. Susan Blackmore, a British psychologist who specializes in skeptical psychological explanations for the paranormal, had her own OBE which prompted her to write a book on the subject, *Beyond the Body*. This was not a Liguori-like marvel: the physical Blackmore remained in the same room the entire time and occasionally described the travels of her mental self to friends who were there with her. They can testify that she never left, there was no one who reported seeing her in any of the places she described having visited, and she herself makes no grandiose claims for her experience, preferring to analyze it psychologically as a hallucinatory episode rather than accept it as a miracle. One such story obviously isn't enough, but there are others, some of them from the most sober of scientists.

One account quoted in a highly technical series on the neuropsychology of vision called *Visual Agnosias* typifies the kind of OBE story that I think supports the idea that people who report OBEs have actually been transported mentally if not physically. An anonymous scientist described how he had been standing in a college classroom delivering a speech to a group of about ten people, when: "I suddenly had the clear impression of observing myself...from a position more than a metre above my head and somewhat to the side; near the ceiling of the room." His experience lasted only about fifteen seconds, during which time his body continued the speech while "he" watched. He was nineteen when it happened, has never had another one, still thinks many of those reported are frauds, and ascribes his to a transitory biochemical effect.

This terse account contrasts with Susan Blackmore's in two important respects. First, hers was a much more elaborate experience: it began with a whirlwind trip through a corridor of trees, then took her around the world, to Paris, New York, South America and the Mediterranean, where she spent time inspecting the trees on a "star-shaped" island, all the time relating her experiences to the friends who were with her. Second, she has gone public. It is a measure of the suspicion that surrounds OBEs that the "well-known American neuroscientist," as he was referred to in the book, remains unidentified even though his account is quite unspectacular.

In the vast territory between Alphonsus Liguori and our anonymous scientist lie thousands of OBE accounts, and from them a typical experience can be pieced together. OBEers are usually in a relaxed state, sometimes just before or after sleep, often very tired, when they suddenly realize that they seem to be floating in the air above their bodies. (The connection to sleep has prompted the suggestion that OBEs are really dreams, but because they require that the person be aware of what's happening, and experience a separation from the body, it's very unlikely that OBEs can be ordinary dreams. However, they are not unlike "lucid" dreams, dreams in which the dreamer realizes he or she is dreaming. More on these in Chapter 19. However even a lucid dream would be a tenuous explanation for our anonymous scientist's experience.)

OBEers can see themselves, other people and the furniture in the room all as they would appear from above. Those who are adventurous enough to suppress panic and stay with the experience find that their floating self has unusual abilities: it can move through floors and walls, visit the homes of friends, however distant, and in some of the weirder cases, visit other worlds. In one such OBE one of the alien residents was heard to say, "Oh, she's from the earth." (It's always so obliging of them to speak our language.) One of the striking features of these experiences is that instead of the overwhelming fear and panic you might expect, many OBEers report nothing but good things: the ability to see with a clarity they have never experienced before, the feeling of an incredible lightness and freedom, the longing to be able to do it again. Needless to say, most find it an unforgettable experience. It's important to note that the people who have these feelings need not have had any preconceptions about the notion of leaving their body, nor are they necessarily spiritually oriented. Much of the disbelief of OBEs stems from the suspicion that those who have them are simply impressionable people who believe so strongly in the existence of a soul or the afterlife that they hallucinate what they most desire. In fact in the 1970s a New York psychiatrist, Jan Ehrenwald, analyzed a set of OBEs and concluded that the common feature was "the perennial quest for immortality" and that this subconscious fear of death prompted the OBE, a short-lived rehearsal of what they hoped would happen when the soul finally escaped the body. In drawing this conclusion Dr. Ehrenwald lumped together the ecstatic "flights" of shamans with the OBEs of people who were comatose with illness. But the skeptical view that the only people who report OBEs are somehow true believers comes less from any psychiatric opinion than from the fact that the out-of-body experience has always been strongly linked to the unusual concept called astral projection.

Astral projection is an idea that comes out of the teachings of The Theosophical Society, founded in the nineteenth century in New York. Theosophy pictures human existence as many-layered, with the physical body being the least interesting and the least essential. Beyond the physical body there are several others, including the astral body, which inhabits an astral world. Theosophists contend that the astral body is capable of separating from the physical body and moving around without it, and because the astral body is the seat of consciousness, its separation causes an out-of-body experience: the body stays behind, while the mind moves with the astral body. If you get good at it, you can launch your astral body from your physical self at will, hence the term astral projection. Obviously an OBE sounds like astral projection, and many of the most enthusiastic proponents of OBEs have been Theosophists.

One of the bizarre subtexts to astral projection is the belief that the astral body is at all times connected to the physical by a silver cord, and some people have reported seeing this cord during their OBEs, including a few who have known nothing beforehand about Theosophy or astral projection. Even Susan Blackmore, who holds no brief for astral projection as an explanation for OBEs, noticed that she was tethered by a silver cord during her experience. Theoretically the strength of the silver cord lessens as the body weakens, meaning that illness or physical exhaustion can set the stage for projection (or an OBE). Freeing the astral body is easier than you might think. Anaesthesia releases it; sleep releases it slightly; even being in a car that stops suddenly may pitch the astral body forward and out of your physical body momentarily (although this has always been even more difficult to claim on your car insurance than whiplash). Once freed, the astral body can move at different velocities, with the top speed allowing virtually instantaneous travel: one moment you're here, the next you're there.

More important than these details is the fact that there are thousands of reports of OBEs, many from those who could be considered the converted, but many not. And as experts on astral projection have pointed out, there are enough coincidences in these reports—even from people who had no idea what to expect—to suggest there is something here worth looking into. One of the most important questions to be asked is: can out-of-body experiences tell us anything at all about the brain? The answer is a lot trickier than the question.

It's difficult to know how common OBEs are, because most

of the surveys that have attempted to answer this question have used atypical groups (like students volunteering for experiments in ESP) or have posed questions that could be interpreted in several ways. However, the numbers of people who claim to have had an OBE range from less than 10 per cent to 34 per cent; as far as I'm concerned any figure in that range is startlingly high. To think that one out of every three people, even in some self-selected group, might have had an out-of-body experience makes me wonder where I've been all this time. As if that weren't enough, people have experienced OBEs in what can only be called mundane circumstances: taking a driving test, giving a sermon (nobody in the congregation noticed anything unusual), standing beside a filing cabinet or even while putting on make-up. One other survey worth mentioning was done by Charles Tart, a psychologist and veteran investigator of the paranormal and the just plain out-of-the-ordinary. He reported in 1971 that 66 out of 150 marijuana users—an amazing 44 per cent—claimed to have had an OBE. Although Susan Blackmore points out in *Beyond the Body* that those numbers don't necessarily indicate a cause-and-effect relationship (it could be that the same people who smoke marijuana happen to be the kind of people who have OBEs) the connection between the experience and a psychoactive drug might be significant. It's also notable that Blackmore's OBE occurred immediately after she smoked some hashish.

Some attempts have been made to capture an OBE in the lab by charting the physiological changes, if any, in a person who is having one. Obviously the experimental subject in this approach must be someone who can call up an OBE practically at will, and even though most people who have an OBE have only one, there is a handful of individuals who are apparently chronic OBEers. Unfortunately there's very little of consequence that has come out of such experiments—brain-wave recordings from even the best and most experienced OBEers are inconclusive. The most that can be said is that when the brain makes the switch from normal to OBE there is no consistent change in the pattern of electrical activity, and sometimes there is no change at all. Most of these studies were done twenty years ago using what was then the best electroencephalographic equipment available, and there's no doubt the same experiments could be done better now. However, given the dubious flavour to this work (at least in the eyes of most scientists who are expert in EEGs) it's unlikely we'll hear of replications of these experiments in the near future. It's too bad they didn't work, because the next step would have been even more interesting: to persuade one of these out-of-body travellers to do the same while being monitored with one of the state-of-the-art brain scanners like a positron emission tomography machine. That would be something: to see which areas of the brain become active during an OBE.

Neither has there been significant success in detecting the presence of the double, something to which the technology of the twentieth century should be equal. After all, if the body that travels in the OBE is so like the original that mourners at the bedside of Pope Clement could be absolutely convinced they were looking at and talking to Alphonsus Liguori (even though the original was lying in a heap four days' journey away) you'd think photomultiplier tubes, electromagnetic field detectors and strain gauges would sense something if an astral body were to land in the room. But so far no such detection attempts have yielded results that seem reliable. This lack of data leaves the field dominated by personal and largely unverifiable accounts of leaving the body, sometimes travelling for hours to distant places, then returning. Not surprisingly the combination of an absence of hard data but a wealth of testimony has encouraged the advancement of several theories to explain OBEs. Some of them pile strange upon strange, like the idea that while something actually does leave the body, that something is a mental creation, not a physical object, and it inhabits, not the real world, but a mental world jointly created and shared by all OBEers. Ideas like this are unlikely to reveal anything about how the brain works.

It's worthwhile taking a breather from the bizarre world of the OBE to approach the topic from another direction: the supposedly tough-minded, objective neuroscientific point of view. Neurologists and psychologists have investigated the idea of people splitting themselves in two, but most of their subjects (or patients) have had, not an OBE, but a vision of themselves: they have seen their double.

There is one immediate and probably significant difference between the OBE and the appearance of a double. If you have an OBE, the subjective "you" is up there above your body. Even if your body carries on doing whatever it was doing, "you" have moved. By contrast, if you see your double, "you" haven't moved at all; a replica of yourself has appeared before you. The striking thing about seeing one's double is the number of writers and poets who have either written about the frightening experience or endured it themselves. Fyodor Dostoevsky, Edgar Allan Poe, Percy Shelley, Wolfgang Goethe and many others knew about and were fascinated by doubles, prompting neurologists to wonder why writers should be so taken with such an uncommon experience.

Those experts who have tried to answer this question by sorting through a mix of literature, diaries, personal testimony and commentary have speculated that there are two or three ingredients necessary—at least among writers—for seeing one's double. All of them were nicely packaged in one person, the nineteenth-century French novelist, essayist and poet Guy de Maupassant. In his later years he saw his double constantly: "Every other time when I return home I see my double. I open the door and see myself sitting in the armchair... Wouldn't you be afraid?" Why was he plagued so? To begin with, de Maupassant had powerful visual imagery, the ability to conjure up vivid scenes in his mind's eye. He was also obsessed with himself, a classic narcissist. The combination of the two was a perfect recipe for seeing a double—what better to see in your mind's eye than yourself, the primary object of your fascination? In addition de Maupassant was likely to have been prone to hallucination because he was in the terminal stages of untreated syphilis. Great powers of visualization, narcissism and some degree of brain disturbance appear again and again among those who wrote about doubles. The problem with conclusions like these is that they are based on second- or third-hand reports and arm's-length psychoanalysis and so may not tell us much that's useful about any brains, even writers'.

However, over the last century and a half neurologists have identified from their own patients some direct influences on the brain that predispose people to see their doubles, including the use of hallucinogenic drugs, high fever, epilepsy, migraines, tumours and even brain damage from carbon monoxide poisoning. It makes sense that some sort of stress on the brain could be found in every case, but what is surprising is that it need not be anything as dramatic as epilepsy or organic brain damage. An example was reported in the late 1950s of a woman who began to see her double immediately after the death of her husband. She returned home from the funeral, opened the door to her bedroom and became aware immediately that there was someone else there: a lady, dressed exactly as she was, standing in front of her and mimicking her every movement. From that first occasion, she was visited by her double every day, and while its image was somewhat indistinct—the lower part of the trunk and legs were unclear—its effect on her was very real: "I am fully aware that my 'double' is only a hallucination. Yet I see it; I hear it; I feel it with all my senses. It is me, split and divided."

Meeting a double ranges from the vivid full-colour experience of seeing, hearing and even touching a representation of one's self, to the mere sense that someone, not necessarily even your double, is there. The latter often happens when people are alone and in danger. There have been many such reports from people who were stranded in lifeboats, but they are even more common among climbers. Mountain climbing probably triggers these experiences because it combines the stress of being in a place where every step may be your last with the additional challenge of too little oxygen reaching the brain. F. S. Smythe, one of the climbers in the unsuccessful 1933 attempt on Mt. Everest reported being so convinced someone was walking just behind him at an altitude of more than 9,000 metres that he stopped at one point, divided into two a piece of mint cake he was carrying and turned to offer one of the pieces to this imaginary companion, only to realize there was no one there: "It was almost a shock to find no one to whom to give it." (At least this companion was benign and comforting—there was an instance in the 1980s where a stranger materialized beside a lone climber and persuaded him to take an alternate route down from the summit, a route that turned out to be much too slow and dangerous, with the result that the deluded climber spent a night alone at -30 degrees and eventually froze some of his toes.) Smythe did not feel that the presence walking behind him was his alter ego, but there have been cases where an unseen presence was clearly felt to be a double, and these hallucinations have as much impact on the person as actually seeing one's self. The great British neurologist McDonald Critchley described a particularly vivid example of a presence: a forty-three-year-old woman who would wake up in the middle of the night knowing there was someone she knew very well in her bedroom; she had no doubt this person was her double. She never actually saw herself, always feeling the presence behind and to her left, but so convinced was she that her double was in the room that she sometimes tiptoed out of her bedroom and around the house hoping to surprise this "other."

Critchley also reported a strange case of what could be viewed as a abortive out-of-body experience, a man who suffered from temporal lobe epilepsy and who felt, just prior to a seizure, that his body has grown to one and one-sixteenth its normal size. He was fearful that his personality would leave his body to enter this extra one-sixteenth and then separate completely, leaving him personality-less. These two cases illustrate the great difficulty facing anyone who would try to locate these strange notions in the brain. The woman who felt her own presence in her bedroom had damage to both parietal lobes, whereas the man with the one-and-one-sixteenth-sized body suffered from seizures that began in his temporal lobes. From the point of view of brain anatomy there is little common ground between the two. Add the evidence that conditions involving many separate parts of the brain like schizophrenia, epilepsy and fever (not to mention narcissism and vivid visual imagery) may precipitate seeing a double, and one can only conclude that seeing one's double happens, but where in the brain it takes place or why is a mystery.

What are the parallels between seeing one's double and having an out-of-body experience? In both, the person's body is seen from an impossible perspective: looking down on or at one's self. Psychiatrist Ronald Siegel at the University of California at Los Angeles argues that this is a common feature of memory: he suggests that it is likely your memories of the last time you swam in the ocean would include images of you as seen from beside or above, perspectives invented by your brain. (Although as the use of the video camera spreads, more and more of what we take to be our memories will mistakenly include events on tape in which we will certainly appear as seen through another's eye.) So the mind's eye is capable of such images under normal circumstances. To take it one step further, Siegel's experiments with people taking measured amounts of hallucinogenic drugs revealed that such bird's-eye visions were commonly induced; remember too that marijuana users reported more OBEs than any other group surveyed, and that hallucinogens have also been implicated in the experience of seeing one's double. Psychologist and OBE expert Susan Blackmore thinks that the OBE, like seeing one's double, requires the ability to create powerful and detailed images that become the hallucinatory material.

On the other hand, there are marked differences between the double and the OBE, one in particular being the emotions associated with each. Those who have seen their double and described the experience have for the most part found it distressing, while as I suggested earlier, people who have OBEs describe the experience as like nothing they've ever felt before: a sense of lightness, freedom and oneness with the universe. Also, while an OBE requires a complete transformation of the visual environment, the double is merely superimposed on the actual background, and in some cases, is even controlled by the person seeing it. The woman who saw her double after returning home from her husband's funeral noted that her double moved in mirror-image fashion with her. Others have reported that the facial expressions of their double change as their own do. Contrast that to an OBE which may take the subject to other lands, real or imaginary.

Both these strange experiences are perfect examples of how much remains to be learned about the brain. It can be said that both OBEs and seeing one's double result when some stress triggers thought processes which blend memories, obsessions, fears and the brain's body image into one fantastic imaginary experience. But so what? Unlike simpler mental acts like recognizing faces, the catalogue of events in the brain underlying these phenomena is simply unknown. And it may never be possible to describe them in the kind of detail brain scientists seek. When mental experiences are of the complexity of these, we may be hitting the wall that some researchers have suggested must be there in the quest for understanding the brain. They wonder if it might turn out that the human brain is not capable of completely understanding itself.

part four FACES

12

Do You Recognize This Face?

In June 1976 the American Viking spacecraft took a routine picture of the surface of the planet Mars at latitude 41 degrees north, longitude 9.5 degrees west. The photo revealed all the standard features of the Martian landscape: craters, plains, irregular rocky mesas and low mountains. It also had something out of the ordinary: a face, or rather, *The Face.* It is a mile-wide mesa with surface undulations that looked like facial features in the light and shadow of the late afternoon sun. NASA scientists saw little of interest in the face, and it remained nothing more than a geological curiosity for four years. Then in 1980 two computer imaging specialists, Vincent Dipietro and Gregory Molenaar, began to take a detailed second look at the face. In the summer of 1981 they reported that their enhancement of the original image suggested that the face was not the result of "natural forces."

That was only the beginning: enthusiasts quickly located a Martian "city" nearby, an apparent complex of pyramids and honeycomb structures located several miles from the face, precisely on a line extended from the face's lipline. The harder the handful of face aficionados scrutinized and processed the photos, the more information they were able to cull from them. An eyeball even miraculously emerged from the shadows of an eyesocket. Science writer Richard Hoagland (or "the former science adviser to Walter Cronkite," as he apparently likes to

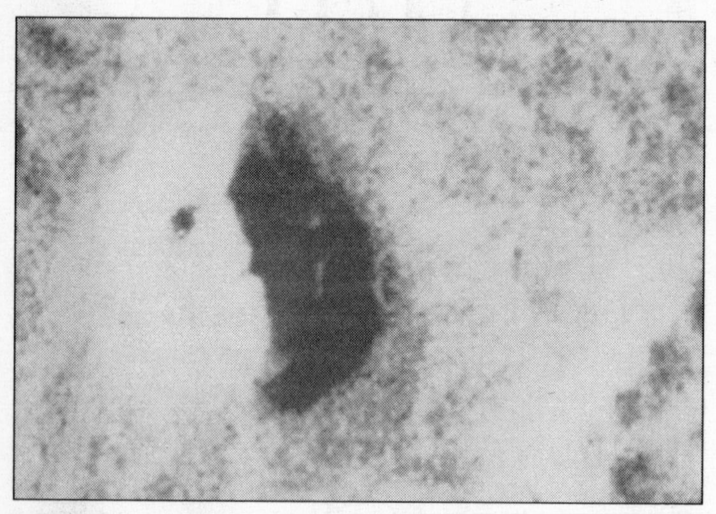

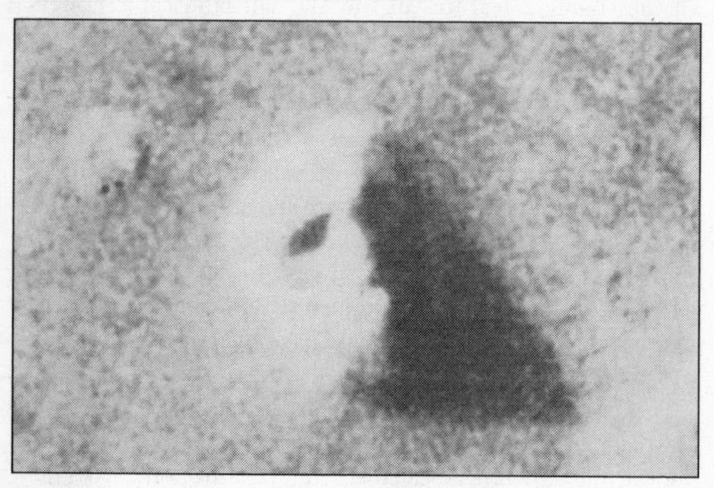

The Face on Mars

be called) even figured out that on the first day of the Martian summer, a being standing in the right place in a complex of hills to the south-west of the face would have seen the sun rise out of the face's mouth. As evidence like this for a face-building culture continued to accumulate, NASA scientists maintained that the face is a table-top mesa no different than hundreds of others on the Martian surface, and that the facial features are simply a fluke juxtaposition of surface irregularities, not a carved monument a hundredfold larger than anything comparable on earth.

The cult of the Martian face came to an absurd climax in the summer of 1993 when NASA lost contact with its Mars Observer spacecraft just as it was supposed to be firing its booster rockets to go into orbit around Mars. A group of face believers reacted by demonstrating outside NASA headquarters, claiming that the spacecraft had been deliberately destroyed by a group of NASA employees who didn't want the reality of the face to be confirmed by the high quality pictures that would have been relayed back to Earth by the Observer spacecraft.

The most interesting thing about this controversy may have nothing at all to do with Mars but everything to do with the fact that the human brain is very good at recognizing faces no matter what the circumstances. It can even turn shadows on a slab of rock into a face if given the chance. Bad lighting, fuzzy focus, a face turned in an unusual pose—none of these deter the human brain from seeing a particular combination of features as a face. There are even parts of the brain which some scientists think are not just skilled at seeing faces, but actually dedicated to that task. It's really no surprise that of the hundreds of thousands of geological features on the surface of Mars there should be one that approximates the right patterns of light and shadow closely enough that any human brain will see it as a face. Once that impression is in place, other brain mechanisms (such as heated imagination) can supply the rest, even to the claim made by one face fan that The Face is aesthetically pleasing and looks "like a king."

The special attention our brains pay to faces is at least as absorbing as speculation about past civilizations on Mars, and has the added advantage of having some evidence to support it. It appears that the brain uses a series of special areas to analyze faces, beginning with the elementary visual components and ending with the name of the person to whom the face belongs. This process is initiated at the back of the brain where visual information is first analyzed, but then with each successive step moves forward, especially along the lower margins of the temporal lobe, the part of the brain around the ears. This sequence of face-analyzing brain areas is especially pronounced on the right side of the brain, so much so that some researchers question whether we bother to use the left side of the brain at all when recognizing a familiar face.

There are people who have lost the ability to recognize faces, even though most or nearly all of their other mental functions are normal. This is as hard to imagine as neglect—people who have this disability called "prosopagnosia" may report that all faces, including those of their family, while they are clearly faces and not teapots, have nothing distinctive about them that would make them identifiable. They're just faces. Others have more serious problems in that they confuse faces with other kinds of objects. There have even been reports of people seeing faces distorted in a way that makes them resemble cubist paintings. But the most common complaint is that faces just don't mean anything any more, and that the only way of identifying someone is to listen for his voice, or his footsteps, or look for some particular piece of jewellery or clothing, or even in rare cases, to study the face for a mole or a scar that would trigger identification. In extreme cases of prosopagnosia an individual who suddenly and unexpectedly finds himself in front of a mirror will think the reflection is a stranger.

I had the opportunity to talk to a man living in Ontario who had suffered two strokes which left him unable to recognize faces, even those of his family. A gregarious man who had been prominent in his community, his prosopagnosia caused him social difficulties. Every conversation with someone on the street had to be prefaced by a comment such as, "I know I know you but I don't remember faces any more..." He told me a story of how he and his wife had gone to Sudbury where she had bought a new blue winter coat. The next Sunday on the way to church she went on ahead, and when he arrived at the church he saw his wife standing talking to the priest. (He knew the priest because of the robes.) After standing beside the two of them for a while he noticed a woman waving at him from a few feet away—his wife. He was standing beside the wrong blue coat. Why didn't he realize the blue-coated woman's voice was not his wife's? At one point he would have been able to, but a second stroke had robbed him of the ability to recognize people by their voices, the usual strategy used by people with prosopagnosia. So he was truly isolated. When I spoke to him on the phone he was looking out the window and telling me that by then he was having trouble seeing people properly as they walked by outside: "A body doesn't really look like a body, just like a cardboard box...I can't see the legs, they don't look like legs, more like a couple of stumps. I can't tell a man from a woman." He had by that time lost more than the recognition of familiar people. He died about a month after I spoke to him.

Identifying the person behind the face can't be the initial step. Your brain first must realize that it is looking at a face and not mistake it for a hat, especially if it is your wife. Much of this early brainwork involves routines which appeared very early in evolution and are found in the brains of all kinds of living things. These include picking out edges and contours from the visual background and combining these features so that real objects stand out. The next step (and the time between each of these is measured in thousandths of a second) is to analyze the features of the object and their relationship to each other to make the decision that it is indeed a face. This step is something special. Even infants a few hours old (or in one experiment, an average of nine minutes of age) show more interest in a circle with simple eyes and a line for a nose than they do in the same circle in which these features are mixed up.

Once the object being viewed is seen to be a face, things really get interesting, because it's at that point that the brain begins to analyze the face for its familiarity. But just how that's done remains unknown. There have been countless studies trying to pin down exactly which features are most important, but there seems to be little consensus. Some have concluded that we rely most on the eyes and hair for identification, but there are other experiments which suggest that the mouth and chin are more important. These two schools of thought were first expressed in the wild west, when some bandits disguised themselves by wearing a Lone Ranger-like mask over the eyes, while others favoured a bandanna covering the nose and mouth. Today robbers unwilling to cast their lot with either face-recognition theory have the option of distorting their entire faces with nylon stockings.

In reality there may be no top ten list of facial features, and apparent dependence on one or the other simply reflects the design of the experiment. Some psychologists have suggested that the overall shape of the face and the relative positions of each of the features to the others is the key to facial recognition, which might explain why the right hemisphere appears to be crucial for face recognition, given that it excels in perceiving the spatial relationships among things.

It is even legitimate to ask if the brain's version of a face is realistic. Why are the extremely distorted versions of famous people's faces—as depicted especially in political cartoons—instantly recognizable? Some of the best bear little resemblance to *any* human face, yet they strike a chord instantly. The chin with the body attached as Brian Mulroney is an obvious example, but other cartoons exaggerate facial features you might never have predicted. Do cartoonists have a special aptitude for getting in touch with their face-recognition neurons? Is the brain's version of a famous face a cartoon in itself?

Several years ago a group of American psychologists used caricatures to try to answer this question. They created an accurate line drawing of a face from a photograph, then used this drawing as the starting point for creating a set of caricatures and "anti-caricatures." A caricature was made by exaggerating the differences between the line-drawn face and an average face; the anti-caricature resulted when these differences were minimized.

Students who then viewed the different faces judged the caricatures to be better likenesses of the person whose face was well known. They also were able to identify caricatures faster than the realistic line drawings. One interpretation of this experiment is that the brain stores cartoon versions of real faces, and so matching these to real faces is both rapid and accurate. If that were true, no wonder cartoons have that instant appeal and ease of recognition—they would represent a more direct route to the stored memory of a face than the face itself. I say "if that were true" because there is still controversy about the meaning of this experiment. I think it is very appealing, because it suggests that cartoonists have unwittingly revealed something of how the brain works.

There are two ways of making even a famous face hard to recognize: either present it upside down or as a photographic negative. The difficulty of recognizing an upside-down face suggests that we do have neural software that is very specifically designed to analyze sets of facial features packaged in a certain way: eyes on top, nose in the middle and mouth below. Obviously there's a tremendous amount of built-in flexibility when it comes to the direction of the face—a profile of JFK is as recognizable as a full-face portrait—but that flexibility doesn't extend to rotating the face 180 degrees. There's some evidence that children don't exhibit this great discrepancy between the abilities to recognize normal and inverted faces until they are about ten years old, suggesting that their face recognition systems aren't mature until then. That suggests that this face recognition system needs an incredible amount of time to be finely tuned, given that infants only days old are already hooked on faces.

Photographic negatives of faces are equally difficult to figure out, and it may be because there is a similar sort of inversion going on in this case, although on a different scale. In 1992 Stuart Anstis, a psychologist at the University of California at San Diego, used himself as a guinea pig for three days in an attempt to determine why negative images are so difficult to interpret. During his waking hours he wore lenses connected to a video camera which reversed black and white and changed colours to their complementary versions. Although he removed the goggles at night, he slept with a blindfold on and showered in the dark. He found that he had great trouble recognizing objects of all kinds, especially faces, and wonders if the fact that our visual systems evolved to perceive things with light coming from above accounts for the difficulty in viewing the world as a negative, the similarity being that anything normally shadowed is bright and vice-versa. If true, then even though the face is oriented normally, a feature like the nose may seem in an odd way to be inverted. Lighting from below is of course the key to the old Hallowe'en trick of holding a flashlight under your chin and turning your face into a horrific oddly coloured mask. And you know what this suggests: there is probably an army of faces on Mars which has so far escaped detection because at the time the Viking pictures were taken the faces were unfortunately lit from the wrong side.

Even monkeys can shed some light on how the brain uses the layout of facial features to identify a face. It has proved possible to record the electrical activity of single brain cells in the monkey's temporal lobes, and many of these cells are sensitive primarily to faces. That the monkey too has devoted a lot of brain power to the recognition of faces, is more evidence that the face is the thing for social animals. Some of these face-oriented brain cells respond to specific facial expressions, some to faces in general, especially in certain orientations, and some to parts of the face like eyes or hair. It's conceivable that these latter cells are tracking the distances between features as well, which would be one way that a familiar face could be stored in the monkey's memory: so-and-so's face is composed of an eye-to-eye distance of 1.5 centimetres, eyes to nose 1.25 centimetres and so on.

However it is done, the brains of both monkeys and humans have an enormous capacity for storing faces in memory and recognizing them almost instantly on sight, even if the previous viewing was years before. The classic human experiment asked people in their seventies and eighties to pick out high-school yearbook faces they remembered out of a mix of such faces, some from their school, some of the same vintage but from other schools. Even those who had been out of high school for fifty years were amazingly accurate.

This study is a reminder that recognition of a face is dependent on memory, which in turn suggests that those who can't recognize faces might have more of a memory problem than a problem of sorting out the face visually. This is partly true: some studies have suggested that people with prosopagnosia, the inability to recognize faces, are nonetheless able to match a single face to its replica included in a group of faces. This would imply that their inability to recognize familiar faces has more to do with not being able to provide a face from memory to match the face they are looking at than with an inability to perceive the face properly. But these experiments are still controversial, and they run counter to the testimony of those with prosopagnosia who claim that as far as they can tell all faces look the same. In one way it doesn't matter—memory obviously plays a crucial role in the recognition of faces, and concrete evidence for that may have been seen in imaging studies of the brain as it sees both familiar and unfamiliar faces.

The late Dr. Justine Sergent at the Montreal Neurological Institute studied face perception and recognition for years, and she described the patterns of brain activity seen when volunteers watched faces appear on a screen in front of them. These images reveal the back-to-front organization of face recognition mentioned earlier. Subjects in these experiments looked at a series of images, each one slightly more complex than the previous, ranging from a blank screen, to simple wave patterns, to faces where the only requirement was to identify the sex, and finally to the complete identification of objects or famous faces (actually labelling faces as actors or politicians). By comparing the images obtained at each step it was possible to show that as the tasks became more complicated, new areas of the brain were recruited.

Sergent showed that the simplest activity of all, looking at a blank screen or wave patterns, activated areas at the back of the brain known to be places where the brain does fairly elementary work on visual images. Gender identification utilized areas farther to the front, and actual face identification activated several areas even farther forward in the brain, most of them in the right hemisphere. One of these was in the temporal lobe adjacent to a structure in the lower part of the brain called the hippocampus. The hippocampus is known to play a critical role in the storage and retrieval of memories. In this case, presumably memories of faces.

The other part of the brain involved in face processing was the front end of the temporal lobe, the final station in this back-to-front analysis of facial images. Both temporal lobes, right and left, become active when volunteers are identifying a given face as an actor or a politician. What's intriguing about this is that the left temporal pole is known to be involved in proper names—if you recognize a face but simply can't name the person, you are likely experiencing a momentary lapse in this part of the brain just under your left temple.

In one of Dr. Sergent's last set of experiments before her untimely death early in 1994, she and her colleagues demonstrated that the left hemisphere *can* recognize a face as long as the identical portrait has been presented before. It is as if the right hemisphere is necessary at the beginning to recognize a face, but once that is done, the knowledge of that face can be transferred to the other hemisphere. But if the same face is presented in a slightly different pose, the left hemisphere is lost. More interesting was an experiment to determine which parts of the brain are active when it is judging the familiarity of a name or a face. A set of names and faces was divided in two: half were actors, half were a variety of professions, including politicians, athletes and singers. Her volunteers were to choose the appropriate group for each person whose name or face they saw. The astounding thing was that for a given person, say Jean Chrétien, the parts of the brain that identify him as a politician from his face are separate from the parts that identify his profession from his name. In other words, Jean—and everyone else—has a double entry in your brain. Why this should be necessary is unknown. There are two views that could be taken. One is that, however redundant this double storage of information appears at first blush, there will likely turn out to be a good reason for it. The other is that the brain evolved in a kind of hodge-podge ad hoc manner, and as a result has been left with certain unavoidable processing inefficiencies. If it were possible to design the brain from scratch, or perform some corporate restructuring, maybe one entry for Jean Chrétien would suffice. The brain hardly ever turns out the way we expect it should.

If instead of faces the brain is asked to identify objects, it goes through much the same sort of back-to-front progression, except mainly in the left hemisphere. Even so there are differences: in Justine Sergent's experiments no object caused anything in the left side of the brain near the hippocampus to become active, which seems to say that the memory processes invoked in face recognition are not the same, or at least don't travel the same routes, as those involved in recognizing objects. Nor was the left temporal pole involved, which isn't really surprising given that it seems to be involved in proper names. You probably don't need it to recognize just any postage stamp, but you might if you had to identify it as an 1851 Canadian twelve-penny black.

The striking thing about these images is the progression they reveal: the trickier the task, the farther away from the primary visual centres it is tackled. Simple images at the back, politicians' faces at the front (although it could be argued that there isn't a lot of difference). The other important observation is that the right hemisphere dominates face perception—this has been a matter of controversy for some years, but these images of the brain at work have resolved much of the ambiguity.

Given this widely distributed set of brain areas that play a role in face perception, it's not surprising that people who have lost the ability to recognize faces make up a pretty heterogeneous group, because damage to any one of these areas will cause a different kind of perceptual disability. There are people who can do no more than recognize that two identical photos are the same face (and some who can't even do that) and others who can see that two faces are the same person, yet won't realize that the two pictures are of themselves.

One of the stranger aspects of all this is the fact that some people who claim that they cannot tell faces apart actually can but are unaware of it. This is called covert knowledge, and it can be demonstrated in a variety of ways. A person who claims not to be able to recognize a single face will nonetheless do better than chance when asked to match names to faces, or conversely will do poorly on memorization tests if they are asked to remember faces paired with incorrect names. Even more bizarre is the demonstration that even though some of these patients won't show a flicker of interest when family members' faces occasionally pop up in a long series of face pictures, their nervous systems do react. A lie detectorlike device will record changes in their so-called skin conductance. Yet when asked, they'll deny any knowledge of any of the faces they have seen. Face recognition can apparently go on without the person being aware of it, showing again that sometimes we don't "know" what's going on in our brains. "Sometimes" might be an understatement—even though the capacity for self-awareness is a feature of the human brain that most scientists would argue distinguishes it from the brains of many other animals, it is also becoming obvious that our brains are highly selective about the information they present for our consideration. Much of what the brain does proceeds unnoticed.

13
Faces, Cows and Upside-down Dogs

In 1988 a Torontonian known in medical literature as C.K. suffered brain injuries as a result of a car accident which left him practically incapable of naming objects. Lab tests reveal that it doesn't matter whether he sees the real thing or a line drawing: in one test he labelled a drawing of a spear of asparagus as "a rose twig with thorns," a tennis racquet as "a fencer's mask" and a dart as a "feather duster." You'll notice that while he only managed to get eighteen out of sixty objects on this test correct (the average score is fifty-seven) he nonetheless had a clear idea of both the general shape and individual features of what he was looking at. A fencer's mask *is* the next best thing to a tennis racquet.

Further tests with C.K. showed that his inability to name things from sight was not because he didn't know anything about them or that he had somehow lost their names. If he handles an object while blindfolded he has no trouble naming it and describing in detail what it is and how it works. So far nothing about C.K. and his naming problems are unique to him—this inability to name, called visual agnosia, is a relatively common result of brain damage. But there are other features of C.K.'s case which make him special.

One is that, while he can't name a clothes peg if you hold it out in front of him, he can draw a clothes peg from memory without any problem at all. Of thirty items that he had failed to name in the line-drawing test mentioned above, he was able to draw twenty-nine accurately, and even well—he studied art in school. However, if you show him a set of his own drawings of objects, as bizarre as it sounds, he will be incapable of recognizing them. Ask him to draw a padlock, and he can do it beautifully; ask him a week later to identify that drawing, and he's likely to call it an earring. C.K.'s ability to create a mental image of an object from its name is intact, but he cannot make the connection between something he's looking at, a visual image, and its name. While it's not completely clear what this says about the brain, there are at least two possibilities. One is that there are two versions of "asparagus" in the brain, one available to be matched to the asparagus spear that you see and another for the spear that you call up in your mind, perhaps in response to the word. It could also mean that there is only one mental representation of asparagus, but different routes for getting to it, one through imagination, the other through vision.

The other oddity about C.K.'s case is something that might shed light on face recognition: while he has suffered this tremendous loss when it comes to naming objects, he has no trouble with faces. Show him a single face, and ask him to identify it in a subsequent set of six, and he can do it easily. He actually scored higher than average on a face-recognition test where some of the faces are seen in three-quarter view or poorly lit. He also had no trouble identifying Imelda Marcos, Boris Yeltsin, Oliver North and even the Queen Mother with a scarf over her head. How can this be explained? The details that would distinguish even two faces as dissimilar as Yeltsin and North are nonetheless pretty subtle compared to the differences between a spear of asparagus and a rose twig with thorns.

The fact that C.K. has serious problems with objects but none with faces has suggested to some researchers that he is living proof that the brain has a special separate face-recognition capacity, a separate software package that deals with faces and nothing else. This is a controversial claim among psychologists in the same way as is Noam Chomsky's argument that there is a specific language organ in the brain's left hemisphere.

You might well ask, who cares? Is there a significant difference between arguing that the brain is especially good at faces and language and going that one step further to argue that each occupies a special place that runs by its own rules? It *is* important in the sense that agreeing that there might be one or two autonomous specialized information processing centres in the brain opens the door to thinking that the entire brain works this way. A number of neuropsychologists think the brain is an assemblage of individual "modules," all of which are co-ordinated to give the impression of a single unified thinking machine. But this is by no means a unanimous view. Much of this argument seems to be driven by faith: either you believe that the brain is a multi-centred association of different specialized processing areas or you don't. If you do, a case like that of C.K. seems to you to be an example of one of those centres.

The counter-argument is that faces are dealt with in exactly the same way as objects (although by different parts of the brain), and one could be damaged while the other is spared. There are plenty of examples that could fall into this category: C.K. can do faces but not objects, while Justine Sergent at the Montreal Neurological Institute once reported a case of a man who, although completely disabled when it came to faces—unable even to match two identical photographs of the same face—was able to give the correct make, model and year of 172 out of 210 pictures of cars, something that no one with an *undamaged* brain has yet been able to do.

To complicate the picture, there are cases of people who lose more than just their recognition of human faces. One woman in Israel who had been an avid birdwatcher could no longer identify birds, even though she could see their colours plainly, hear them and follow their movements. She simply said, "All the birds look the same." Another patient, a farmer, had what was probably a stroke when he was fifty-eight, and when he regained consciousness was unable to recognize his doctor, his neighbours, even his wife and brother. Gradually his recognition of those closest to him returned, and no other problems were identified until he returned to his farm. There he found to his dismay that he was no longer capable of what had been his most prized ability: recognizing individual cows and calves. He had been able to differentiate them by their facial expressions, but now, sadly, every cow looked the same. The man seemed more distressed by this loss than his continuing inability to recognize some people. And understandably so: learning to differentiate cows' faces is acknowledged to be a very difficult and usually unrewarding task. Most artists apparently fail if they attempt to depict individual animal faces, although a Polish painter named Kossak once painted a herd of horses in which each horse had its own facial expression.

So what does all this mean? If there is a special brain software package for human faces (which comes loaded in the computer) must there also be others for the entire animal kingdom? Or does the fact that a few individuals are capable of recognizing different cows' faces or birds or stamps simply mean that it's possible with a lot of hard work to learn to discriminate among very similar things?

Some researchers have picked up on that idea of hard work and repetition and argued that psychologists can be led astray into thinking that the brain processes faces differently from anything else because they spend most of their time studying people who have lost the ability to recognize faces. These skeptics claim that the human brain can acquire an expertise in identifying anything; it's just that face recognition is a skill shared by all humans, while identifying birds, automobiles or ancient Greek amphorae is extremely rare. So the chances of finding cases where brain damage interferes with face recognition are much greater than the likelihood of finding a farmer who can no longer identify his cow's faces, simply because the population at risk for such a disability is huge. Every human is an expert in faces; one in a million becomes expert in the recognition of something else.

Rhea Diamond and Susan Carey at the Massachusetts Institute of Technology tried to prove this point by comparing peoples' abilities to recognize dogs (the whole animal) and human faces, both right side up and upside down. They used both orientations because it's well known that our ability to recognize faces is poor if the faces are upside down. Carey and Diamond reasoned that if face recognition is just one example of expert knowledge, then dog experts might well experience the same dramatic reduction in their ability to tell individual dogs apart if the pictures of the dogs were upside down. They chose pictures of Irish setters, Scottish terriers and poodles, ensured that factors like the animal's stance and the backgrounds were uniform, then presented these photographs to experts from the American Kennel Club Directory. Each expert saw a series of dog pictures, then saw each of those dogs paired with a similar one: their job was to pick out the dog they had seen before. The series were presented both right side up and upside down. Diamond and Carey discovered that upside-down dogs were much harder for these experts to recognize. It sounds tough to me—I could see a line-up of Scottish terriers and not be able to tell them apart—but these people were breeders, handlers and judges. They knew their dogs (as long as they were right side up). It might sound paradoxical that those people who have scrutinized setters and poodles for decades are just the ones who are thrown off the most by seeing them upside down, but that is the point this experiment was trying to make. These are likely the only people who can recognize dogs well enough right side up to demonstrate the same dramatic loss of recognition that the rest of us show with faces, our subject of expertise, when *they* are turned upside down. Does this experiment prove that face recognition, rather than being an ability with its own dedicated brain circuitry, is nothing more than the kind of expert recognition that can be developed for dogs, stamps or chandeliers? It suggests that might be true, although the fact that we seem to be unusually interested in faces practically from birth is not addressed by this experiment. The brain does faces automatically—it has to learn to do poodles. And it's also not likely that neuropsychologists who believe that the brain has a face processor will be dissuaded from that view by some upside-down terriers.

We aren't the only species whose brains do a very good job of reacting to faces. Monkeys' brains do too, and experiments have established that they react to both human and monkey faces, and even that there are individual brain cells that fire when a face is seen. The fact that monkeys' brains respond to faces didn't really come as a surprise, because monkeys are highly social animals which use a variety of expressions to communicate anger, fear and threats. The brain of such an animal would be well advised to have a special sensitivity to the face on which such signals are displayed. (And most scientists would also accept the idea that our face-recognition abilities trace their existence back to a common ancestor of monkeys, apes and humans.) But few researchers had assumed that another familiar animal would also show a highly developed ability to pick a face out of a crowd: the sheep.

The sheep in question were suspended in a hammock facing a screen a metre in front of them. Life-size images appeared on the screen for five seconds at a time, and while the experimenters didn't try to control the sheeps' eye movements, they did note that it seemed likely the sheep were looking at the slides, because whenever there was a picture of another sheep, or a bowl of food, they bleated. Electrodes implanted in the sheeps' brains recorded brain-cell activity in response to four different sorts of pictures: the majority fired when the animal in the hammock saw a picture of a horned sheep, like a Mouflon, especially one with large horns; a second kind of cell responded to familiar sheep as opposed to ones the subject had never met; some brain cells fired in response only to sheepdogs and humans; and a last group of cells responded to a number of different kinds of faces, including pigs. Interestingly these cells failed to fire when upside-down faces were presented, reminiscent of the inability of humans to recognize upside-down faces. (Cells in monkey brains do respond to upside-down monkey faces, but the argument there is that monkeys often see each other's faces in that position. Sheep rarely see upside-down sheep faces.) The fact that some of these cells tailor their response to the size of the horns suggests that social factors play a key role in the deployment of these cells. A sheep with a brain that can size up relevant status signals quickly is a sheep that is likely to live to fight another day.

It's also fascinating that there are cells which react to sheepdogs and humans, because these are relatively recent intruders into the lives of sheep. The domestication of animals happened only a few thousand years ago—the sheep's brain has been evolving (although not with any great speed) for millions of years. This is not unlike the apparent paradox that the human brain has developed systems for reading, another activity only a few thousand years old. In both cases the most reasonable explanation must be that there were systems already in place with similar capacities to those required by these "new" abilities, and these systems simply adapted to the greatest need. As this experiment showed, there are cells in the sheep's brain which respond to faces in general—maybe a group of cells like that were forced by domestication to tune to dogs and people.

I can't leave the subject of face recognition without relating a couple of strange tales in which it's clear that the brain, whether it has a special face module or not, can still treat faces as unique objects. A case reported by Dr. Otto-Joachim Grüsser involved a woman out for a walk with her husband when she suddenly noticed that the left side of his face had become distorted and swollen. Her immediate panic subsided when he claimed he felt perfectly normal, but they hurried home to look at themselves in the mirror. Of course they both looked normal to him, but she saw not only the horrible transformation of his face, but also of her own. For about two weeks she continued to see gross distortions on the left sides of faces, even those in photographs, although in every other way her perception of facial expressions, and identification of famous faces, was normal.

In another case reported by Dr. Grüsser, a sixty-two-year-old woman had had a stroke, recovered, and nine months later was sitting on a subway train, across the aisle from a passenger holding a poodle in her lap. When she glanced up, she saw to her horror that this passenger had a poodle's head. We've all heard that dog owners begin to look like their pets after a while, but this was something else: a serious, and unsettling, face distortion. Every other passenger on that subway train appeared to the woman to have a poodle's head too, as did every person she encountered after she fled the subway. The whole weird experience lasted about half an hour. This woman had a second experience at work, where suddenly all the other office workers appeared to have a series of eyes and noses ringing their heads whenever they looked one way or the other. There is really no explanation of these cases, other than to say that it is no surprise that these kinds of odd hallucinations should involve the face, the one object that we devote more time and brain space to than any other.

And finally, the age-old problem: Why is it so difficult to come up with the name for the face you know so well? One suggestion is that the name is the last in a series of steps that occur when you see a face. First the face is identified as a face, then it's assigned a gender, recognized as familiar and connected with all the other information you possess about the person. Only then is the name added, the neurophysiological equivalent of the cherry on the sundae. Because the name is the last step, it is in some sense the most vulnerable. This theory also predicts that you would never come up with the name without already knowing everything about the person; you won't look at a face, say "That's Leonard Cohen," and then be unable to remember anything about him.

As is the case with absolutely everything about the brain, there are competing theories. One argues that names are unique, whereas each of the other single attributes of a person is likely shared by others. So there are thousands of pop singers, as many poets, several from Montreal, but only one (at least in this context) Leonard Cohen. The more specific the information (i.e., the name) the harder it is to retrieve it. Another idea, suggested by Gillian Cohen (no relation) of the Open University in England, is that names are hard to remember because they don't mean anything. For example, Ingram means nothing (even to me), whereas writer does, and so writer is easier to remember. Gillian Cohen made up a list of names and occupations where these usual roles were reversed: the name meant something but the profession didn't. These included Mr. Baker the ryman, Mr. Cook the rimmer, Mr. Barber the sloman and Mr. Porter the walden. This time around people remembered the name rather than the job, the opposite of what usually happens, suggesting that the meaninglessness of names may contribute to our inability to remember them. They do sound like interesting jobs, though.

None of these theories seems to explain what most of us have experienced from time to time: the importance of context. A face itself isn't enough, because if you see it in an unusual context you're likely to have to struggle to place it where it belongs before you are able to identify it. And this is not an informational context, but a visual one. If you see your veterinarian shopping, it takes a while to realize who it is. Imagine that face framed by a lab coat with the smell of disinfectant in the air, and you'll get it.

14
The One-Sided Face

If you took the time to list the number of different places in the brain that apparently play some role in perceiving or analyzing faces, you'd have to conclude that "doing" faces is the brain's *raison d'être*. Right side, left side, front and back; all have been shown to be involved. Presumably this is because we are social animals living in situations where it is of paramount importance to recognize those you know, and more important, to know how they're feeling. Nothing in the last two chapters addresses this aspect of face perception: reading the expression on the face.

This ability is not uniquely human. The great apes—to whom we are most closely related—and even monkeys manage their social relationships by a variety of displays, many of which use the face for their expression. And although it might make those who don't believe in evolution uncomfortable, there are some obvious parallels between the facial expressions of monkeys and men, especially those of fear and anger. But given that the size of human social groups (and therefore the number of faces that any individual comes in contact with) is much larger than those of other primates, the demands on the human brain's capacity for recognizing and remembering faces—and deciphering the messages those faces are sending—must be huge.

There is pretty good circumstantial evidence that the ability to perceive facial expression evolved a long time ago, and therefore is to a certain extent hard-wired and not simply learned. The fact that monkeys have areas of their brains dedicated to faces and that the understanding of facial expressions appears very early in human infancy (certainly in the first few weeks, if not days) suggest this is work that the brain has been doing for at least hundreds of thousands of years.

Another piece of evidence in favour of the innateness of facial expressions is that they are the same across cultures. This has been established over the last two decades, largely through the efforts of Paul Ekman at the University of California at San Francisco. Pictures of faces expressing six fundamental human feelings: surprise, anger, happiness, disgust, fear and sadness shown to a variety of Western and non-Western cultures revealed that all these people understood and agreed to the meaning of the expressions. Ekman went to some lengths to ensure that the people whose reactions he was seeking had not been previously influenced by outside sources of information. Fearing that some of the non-Western groups he had tested might have been exposed to television programs revealing Western interpretations of facial expressions, he sought out cultures in New Guinea who had never seen television. The only difference he found was that these people did not distinguish fear from surprise. The consistency of findings also held true when the New Guineans posed the facial expressions for interpretation by Westerners. While there are anthropologists who still maintain that facial expressions are culturally determined, and there are details that are still unresolved (such as whether these six represent the total number of universal facial expressions), Ekman's research is nonetheless widely accepted.

If it's true that there is a fundamental set of innate human facial expressions which are consistent regardless of cultural influences, then these expressions can be seen simply as signals which allow one brain to communicate directly with another. If you experience a certain emotion, it is possible for me to empathize with you without a single word being exchanged. Facial expressions are still a very powerful mode of communication in a largely verbal world, and must have been even more important in the days before spoken language when gestures, sounds and expressions were everything. None of this is very startling—it was only when researchers looked *into* the brain that the surprises started to come. It turned out that, just as in the case of recognizing a face, there are parts of the right side of the brain that are specialized for interpreting facial expressions.

One way to demonstrate the right hemisphere's ability to perceive facial expressions is to flash pictures of expressive faces briefly off to one side or the other while a person stares straight ahead. Images of faces appearing off to the left will be sent first to the right hemisphere, faces on the right to the left. This pattern follows the general crossing-over of sensory information on its way to the brain. Studies like this have shown that recognition of the facial expression is superior when the image arrives in the right hemisphere. For instance, when shown a cartoon face with a carefully drawn expression for less than a tenth of a second, followed by a second face for a longer period, most people can determine more accurately whether the expressions on those faces were the same or not, if the faces appeared off to the left, in their so-called left visual field.

This makes sense given that the right hemisphere seems to have the predominant role in recognizing faces—it's not such a big jump to go from the realization that you are looking at a face to analyzing the expression on that face, so it would be economical to recruit brain cells for both jobs from areas next to each other. There is good evidence from stroke victims that facial expression analysis is carried out on the right side of the brain close to, and maybe even overlapping, the areas for face recognition. But the fact that there are patients that have lost one but retained the other shows that the two areas are different.*

But the right hemisphere's role in facial expression doesn't end with perceiving them—it also plays a dominant role in producing them. As odd as it may seem, study after study has concluded that the left side of the face expresses emotions more strongly than the right side (this is true of monkeys as well as humans), and it is the right hemisphere which controls those more active left-side facial muscles. One simple way to demonstrate this is to divide the photograph of a face down the middle into separate left and right sides, then make mirror images of each of the sides and use them to create two new faces. One would be the original right side plus its mirror image, the other the original left and its mirror image. Whenever this is done with faces portraying a variety of emotions, neutral observers judge the original left-side plus mirror image to be the more expressive. It's easy to imagine even if you have never seen it: if in the original smiling face the left side of the mouth was turned up farther than the right, then the left-side plus mirror image face would have a mouth turned up at both corners, while the right-side plus mirror image would have a mouth that was practically flat.

Why would we have asymmetrical facial expressions? Given our tremendous facial musculature and the importance of facial expressions, it seems a shame to limit their scope by

* There has even been the occasional patient who can recognize facial expressions perfectly well but has lost the ability to name them. One in particular was able to choose the appropriate facial expression to apply to a scene, like a funeral, or to a sentence read with an obvious emotional tone; he could tell whether two faces were expressing the same emotion or not, but when it came to labelling a particular expression as happy, sad or angry, he just couldn't get them right. And yet when this patient was presented with the same sort of problem expressed verbally ("name the emotion on the face of a person who is being attacked by a vicious dog") he labelled every one correctly. He knows expressions and he knows the words for those expressions, but he can't connect the two.

playing them out largely on one side of the face. I don't think there's any answer that a majority of experts would accept, but there is one that takes into account the right hemisphere's dual responsibilities of perception and production of facial expressions. Imagine you are looking at a face directly in front of you. The image of the right side of that face (which is on your left) arrives first at your right hemisphere, and vice-versa. You will recall that the right hemisphere is better—or at least quicker—at recognizing facial expression. The irony is that the expression that you are trying to interpret is being portrayed more vividly on the left side of the face, and therefore arrives at your left hemisphere, the unskilled one when it comes to faces. So what gives? Why would one brain's strengths play to another's weaknesses? One simple explanation is that the left side of the face is forced to exaggerate its expression *because* the hemisphere that is perceiving it (your left) has to have things underlined and capitalized to catch on. The subtler expression on the right side of the face is perceived easily by your more skilled right hemisphere.

This speculation is somewhere between a hypothesis and a guess. Others have suggested that this entire right-side, left-side disparity is irrelevant in the real world because when you look into someone's face you are almost never looking straight at it anyway. Your eyes play over the surface of the face you are looking at, perhaps glancing first at the left eye, then at the corner of the mouth. In addition, neither the expression on the face nor the face itself remains still for more than a fraction of a second and the brain is processing the changing expression information on the fly. This skepticism sounds reasonable, but there are experiments that contradict it by showing that, at least in the laboratory, people asked to scrutinize a face closely spend about two-thirds of their time looking at the face's right side. Does that mean that we do the same thing in everyday life? I don't think anyone can answer that.

This research has taken off in some odd directions. One is the discovery by Dr. Otto-Joachim Grüsser that the vast majority of a set of hundreds of portrait paintings executed from the 1400s through the 1600s depicted the head turned towards the right shoulder. This of course exposes the more expressive left side of the face. Perhaps the most intriguing aspect of this is that the practice gradually declined from the 1700s to the present, although at all times it was more pronounced in female portraits than male. You could easily argue that this was no more than an artistic convention, but it is still worth asking how it started and whether there was some unspoken recognition that the left side of the face was special.

There is an odd parallel here to one of the great unsolved questions of human behaviour: why do women hold their babies with the babies' heads on the left? There is no doubt that they do: observations of women in a variety of cultures, surveys of photographs of Western, Eastern and North American native cultures and experiments all agree that about 80 per cent of women cradle babies in their left arm. (This is in contrast to the way people hold bags of groceries. I. Hyman Weiland of the University of Southern California observed that half of shoppers carrying bags about the size of babies held them on the left, and half on the right.) Most people will assert that because most women are right-handed, holding the baby in the left arm frees their dominant hand to do other things. This unfortunately does not explain why most left-handed mothers also hold their babies on the left. In the early 1970s psychologist Lee Salk suggested that mothers held babies this way so that the baby could hear the mother's heartbeat, a familiar and comforting sound from the womb. Because the heart angles towards the left of the body, it would then make sense to position the baby's head to the left. Salk was even able to show that babies in nurseries who heard a recorded heartbeat over a loudspeaker gained more weight over a four-day period and cried less than babies who did not hear the recording.

But scientists have continued to wonder whether this is the whole story, and more recently some studies have suggested that mothers hold babies' heads on the left for reasons that might have more to do with the brain—and in particular the right hemisphere—than the heart. Imagine cradling a baby and looking down at the baby's face. If it's on your left, then the face is in your left visual field, and a quick glance ensures that the image of that face will arrive at your right hemisphere, where the emotional expression on the baby's face will be read swiftly and surely. In addition, a baby held in that position sees the left side of your face, and so gets the benefit of seeing the fullest expression of your emotions. Dr. John Manning of the University of Liverpool has shown that mothers who wear eye patches over their left eyes (and therefore wouldn't get the benefit of having the image of the baby's face sent to their right hemispheres) have a much reduced tendency to left-side cradling. While this can't be attributed with certainty to the brain hemisphere theory, it at least suggests that heartbeat isn't the entire answer.

Researchers have even done for baby-holding what has been done for the presentation of the left side of the face in portraits. They have compiled several centuries' worth of paintings and sculptures of mothers holding babies and there is the same mysterious inconsistency as there was for the portraits. If you combine the results of several studies, it seems as if left-side holding was more common until about the late 1400s, then right-side holding gained ground. The two sides remained about equal until the left side started to regain its supremacy in the 1600s, and maintained its lead until today. Given the obvious pitfalls of depending on art to reflect exactly what was happening in society, this waxing and waning of left-side cradling through the centuries demands some sort of explanation. If the paintings are telling it like it was, and women were really changing their baby-holding habits, the sound of the heartbeat theory seems even less tenable. If left-side cradling is preferred because it links mother's and baby's right hemispheres, then psychologists are faced with accepting that some sort of weird alteration in brain organization happened in the fifteenth and sixteenth centuries. The situation is even more complicated than this, because a survey of sculptures from Central and South America has revealed that there was only one period, from AD 300 to 600, and one place, Central America, where left-side holding was predominant.

One way to interpret these findings is to accept them at face value, and connect them to the brain theory of left-side holding by arguing that the differences between the two brain hemispheres can change from time to time as a result of cultural forces. By this reading, something happened in Central America between AD 300 and 600 that exaggerated the left-right differences and so made left-side cradling predominant, while during the same period in the cultures in the Andes nothing comparable happened. In Europe, some mysterious cultural force reduced the need for left-side cradling for two centuries or more, then gradually disappeared. There is some evidence that the degree of difference between the hemispheres can be altered by social factors, but this interpretation is still a stretch. The easier alternative is to dismiss it by saying that the artistic representations are not accurate records of what was happening and leave it at that.

The perceptive reader will have noticed that I have made no mention of men holding babies, and with good reason. Men show no preference for left-side holding. Even exhaustive searches of family albums suggest that men position the baby's head randomly, right or left. Is this because the sound of a male's heartbeat is not reassuring to a baby? That seems unlikely. Is it because males are not as good, or to put it more precisely, not as right-hemisphere specialized for detecting emotional expressions on faces? There is some evidence that supports this, and for the moment that is probably the best explanation for the fact that women prefer to hold babies head-to-the-left and men do not.

I can't leave the subject of faces without mentioning some of the strangest research of all. It seems that not only does the experience of certain emotions create facial expressions to match (you're happy, you smile), but the reverse is true as well:

smile, and you will start to feel happy. This bizarre notion comes from research begun in the 1980s by Paul Ekman, and has been elaborated on by many other researchers. Ekman asked volunteers to make the faces associated with certain emotions, without asking them to "smile" or "frown" or "scowl." Instead, a typical set of instructions would be, "Pull your eyebrows down and together, raise your upper eyelid and tighten your lower eyelid, narrow your lips and press them together." When subjects followed these tricky instructions (which produce an angry face) there was an increase in heart rate and electrical conductivity of the skin. Such changes associated with the negative emotions of anger, fear and disgust were consistently greater than changes accompanying a happy face, and were the kinds of changes that would be seen if the person were actually experiencing the emotion that goes with that face. The instructions were given as a list of muscle-by-muscle movements to disguise the emotion subjects would end up displaying, in the hope that they would then be less likely to influence their own bodily reactions. Although it's obviously possible that as you follow the instructions above it might become clear that you are making an angry face, Ekman and his colleagues have found that people's recognition of what emotion ends up being displayed on their face is surprisingly poor.

Ekman himself stops short of saying that by simply setting your facial muscles in the right position you can experience the emotion that would normally be associated with that expression. It's clear that you can cause changes in your body that are normally found alongside that emotion, but that's not quite the same as experiencing the emotion. However, there are some suggestions that volunteers in these experiments did experience some emotional changes, especially with the expressions for negative feelings.

How is it possible that by making the face associated with a particular expression, you can cause changes in your physiology, if not experience the emotion itself? Ekman suggests that there is a network in the brain that controls the entire set of reactions accompanying a strong emotion. So when you are frightened by an unexpected sound in the dark, your face takes on the fear expression, your heartbeat quickens, hormones are pumped out, blood is shunted to the muscles to prepare to flee or fight and you may even shout involuntarily—all these reactions are produced by interconnected circuits of brain cells. Paul Ekman argues that this emotion network can be activated even in the absence of the usual cause (in this case the unexpected sound) if any one of the effects, like the facial expression of fear, is deliberately created. As long as one station is activated the entire network turns on.

Ekman's is not the only explanation of this unexpected phenomenon. Others have argued that rather than being confined to cells in the brain, this effect is a much more roundabout process: the brain commands certain facial muscles to contract, their contraction squeezes some blood vessels leading to or exiting from the brain (while leaving others alone), and that altered pattern of circulation changes the activity of the brain in some regions but not others, the ultimate effect being a change in feeling. Arguments about the actual mechanism aside, the very idea that the smile may precede the happy feeling, rather than the other way around, is a perfect illustration of how nothing about the brain can be taken for granted. Are jokes funnier because we start to smile in anticipation halfway through, and by so doing elevate our mood just in time for the punch line? Does that explain why a well-known comedian gets laughs even on nights when the jokes aren't funny?

part five
MEMORY

15
H.M.

On August 25, 1953 a twenty-seven-year-old man in Hartford, Connecticut and his parents agreed to a last-resort treatment to alleviate the epileptic seizures that were ruining his life. He didn't know then that the operation would make him the single most written-about medical subject in history. The tragedy is that even now he doesn't know it.

He is known in the medical journals and texts only as H.M. In 1953 H.M. was suffering an average of ten petit mal seizures daily, and although he didn't lose consciousness during these half-minute episodes, he would lose contact with those around him, involuntarily closing his eyes and crossing his arms and legs. These "minor" seizures left him in a state of near-constant confusion. About once a week H.M. also suffered a much more serious grand mal seizure which caused him to convulse, lose consciousness and occasionally injure himself. Drugs at the highest tolerable doses had failed to control either kind of seizure, and brain surgery was considered as the final alternative. This is not as extreme as it might sound: surgery to alleviate seizures is still practised today, and many epileptic patients have had their lives transformed by the removal of a small piece of brain tissue, usually a piece containing a scarred or damaged area. Although the electrical storm of a seizure may sweep out of control across the brain, there may be a single trigger point. By eliminating that damaged tissue, there's a reasonable hope that the seizures will stop.

H.M.'s case was not perfectly clear-cut. EEG recordings showed no specific seizure source, just vague disturbances on both sides of the brain in the areas called the temporal lobes. In addition, the only hint that H.M. might ever have suffered the kind of brain damage capable of initiating seizures was the fact that he had been knocked unconscious for about five minutes in a collision with a bicycle when he was just seven years old. He had his first minor seizure when he was ten, and his first major seizure, in which he lost consciousness, at sixteen. H.M. managed to graduate from high school and work in relatively undemanding jobs until finally the seizures made even those too difficult. After every conceivable drug had been tried without success, H.M. became a candidate for surgery. A prominent brain surgeon, Dr. William Beecher Scoville, was to perform the operation, and he was forced to make a difficult judgement. On the one hand there was the equivocal EEG record which offered him no clear-cut target for surgery; on the other, the fact that H.M. was incapacitated by his seizures. Dr. Scoville decided to cut away substantial pieces from the temporal lobes on both sides of H.M.'s brain. Although this particular surgical approach was acknowledged by Scoville to be "frankly experimental," this was, after all, the golden age of psychosurgery: a variety of techniques for removing pieces of the brain, large and small, had been developed both to pacify psychotic patients and curb the seizures of epileptics. So while "frankly experimental" suggests that Dr. Scoville might have embarked on this treatment with a great deal of trepidation, there was much more acceptance of "cut and cure" at that time than there is today.

The surgery was uneventful—in fact, H.M. was awake and

talking throughout. Dr. Scoville removed about eight centimetres (a little more than three inches) from the inside surface of both temporal lobes. This included not only some cerebral cortical tissue but also parts of what is called the limbic system, a collection of several different structures which underlie the cortex and are responsible for emotion. In H.M.'s case, William Scoville removed most of the hippocampus and all of the amygdala from both sides of the brain. The hippocampus is so named because it is supposed to look like a pair of seahorses, but for my money it's a wishbone, joined near the back of the brain with the two arms extending forward, tucked inside the temporal lobes. Attached to the tips of the hippocampus are the amygdalas, small almond-shaped (hence their name) bulbs. The limbic system to which both the hippocampus and the amygdala belong has also been called the "mammalian" brain, meaning that it is ancient in the evolutionary sense, and can be found even in the brains of animals not known for their thinking prowess, like gophers, opossums or even my cat.

The tragedy of H.M.'s story can be traced to the fact that in 1953 little was known about the function of these parts of the brain. One of the first public hints that something had gone terribly wrong came from William Scoville himself, writing in *The Journal of Neurosurgery* in 1954. In summing up a series of surgical treatments for both epilepsy and psychosis, he stated that removing tissue from both sides of the brain in the area of the temporal lobe "has resulted in no marked physiologic or behavioral changes with the one exception of *a very grave recent memory loss* [italics his]...." That it was a time of exploration, experimentation and misplaced optimism cannot be doubted: Scoville muses at the end of this article about the possibility that one day brain surgery for epilepsy might replace the use of drugs completely.

We now know, largely because of H.M., that the hippocampus is responsible for laying down new memories. The removal of the bulk of both arms of his hippocampus left him amnesic: he has made no new memories, except for the most fleeting

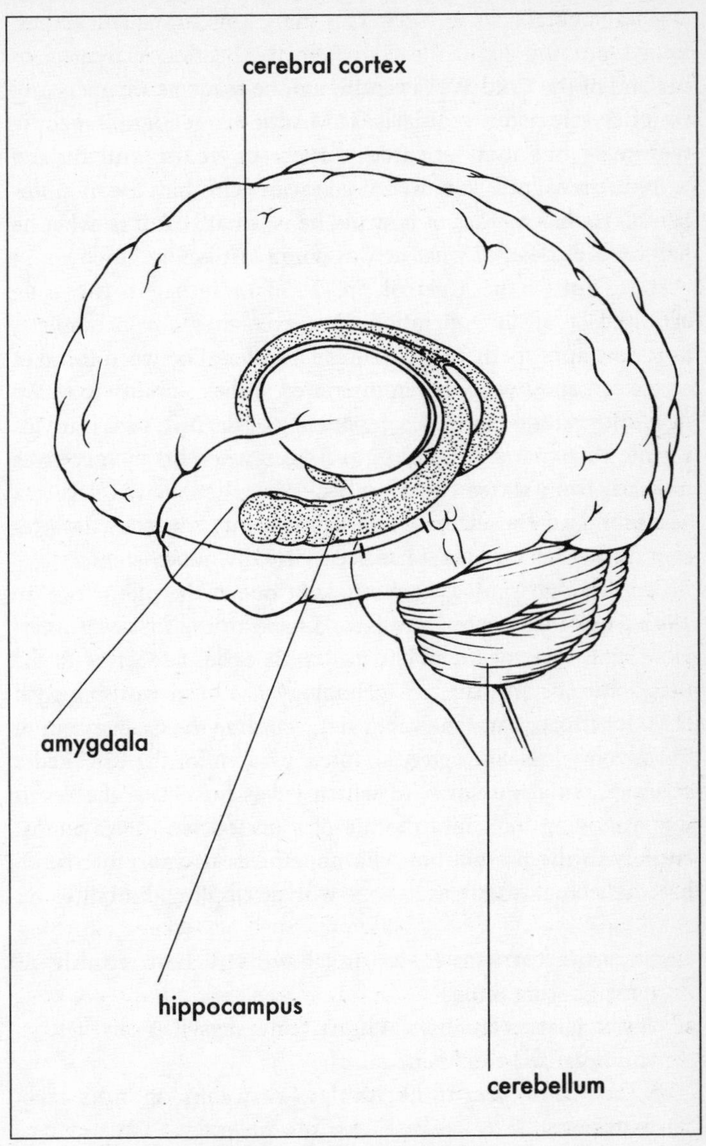

The hippocampus showing the extent of both moderate and radical removal

and fragmentary, since 1953. This man, now in his late sixties, knows nothing about the Gulf War, the Beatles, Watergate or the end of the Cold War, even though he reads newspapers and watches television regularly. H.M. doesn't recognize people whom he has met repeatedly over the years, and indeed wouldn't recognize you even if you were with him five minutes before. He has no idea of how old he is, what year it is, what he had for breakfast, or what he was doing just before lunch.

H.M. does have a set of pre-1953 memories—those he acquired before his operation—but even they are incomplete, as it now appears that the seizures he suffered between the ages of sixteen and twenty-seven interfered with his ability to retain new information. So H.M.'s experience of his own past is divided into three parts: from birth to sixteen when his memory was normal, from sixteen to twenty-seven when it was severely compromised by seizures, and from twenty-seven to the present during which time it has been virtually nonexistent.

Even though H.M. is carefully kept out of the public eye, in 1992 I was lucky enough to have a hand in the first-ever interview H.M. gave for the public media. Dr. Suzanne Corkin of the Massachusetts Institute of Technology has been working with H.M. for thirty years, analyzing and recording the exact extent of his memory loss. She agreed to interview him for the CBC Radio series "Cranial Pursuits" of which I was host, and the result was a moving look into the life of a person who lives almost entirely in the present but who nonetheless tries to martial all his intellectual resources to cope with a crippling disability:

> Suzanne Corkin: How long have you had trouble remembering things?
> H.M.: That... (laughs)... I don't know myself. I can't tell you because I can't remember.
> S.C.: But do you think it's days or weeks, months... years?
> H.M.: Well, you see, I can't put it exactly on a day, week or month... or year basis.

S.C.: But do you think it's been more than a year that you've had this problem?
H.M.: I think it's about that. About a year or more. 'Cause I believe I had... this is just a thought that I'm having myself... that, well, possibly I had an operation or something.
S.C.: Tell me about that.
H.M.: I don't remember just where it was done. And unhhmm...
S.C.: Do you remember your doctor's name?
H.M.: No, I don't.
S.C.: Does the name Dr. Scoville sound familiar?
H.M.: (Leaping in immediately) Yes that does.
S.C.: Did you ever meet him?
H.M.: Yes I think I did. I think I met him in his office.
S.C.: And where was that?
H.M.: Well, I think of Hartford right away. (pause) Well, what he learned about me helped others too and I'm glad about that.
S.C.: How do you feel about having a problem remembering things?
H.M.: Well, I don't feel good about it, but ah I... what is found out about me will help the medical people [with] other people too and help them.

When H.M.'s thoughts turn to his childhood, he is able to recall the same sorts of events that anyone in his late sixties might, although Suzanne Corkin has pointed out there is one difference. The memories we have of our childhood are constantly being revised and refurbished. When my brother recently sent me a set of photographs of a Christmas in our house many years ago, there I was sitting in front of the tree holding a newly opened toy rocket. I had always believed that I bought it in the summer with money from painting the backyard fence, but now I know how I acquired it. H.M. has no such opportunity, because even if presented with a photo album, he would forget it moments later. So the childhood memories he is able to relate are in their original form, and Suzanne Corkin fears that as time passes they are becoming more tattered and fragmentary as a result:

> "What is the favourite memory you have of your mother?"
> "...That's she just my mother."
> "Do you have a favourite memory of your father?"
> "Well..."
> "Can you remember a special event with your father, a trip or something like that?"
> "Well, there I'm stuck in a way because I remember a trip and I remember something else too. One is the trip that goes over the Mohawk trail 'cause we've been over it so many times... That was my mother and my father and myself."
> "Can you tell me a little more about that?"
> "It was near Jacob's Ladder and there's a hairpin turn at Jacob's Ladder."
> "What year was that trip?"
> "Gee I've forgotten how many times I've been over it."
> "Was it recently or a long time ago?"
> "No it was a long time ago I'd say."
> "Tell me about your friends."
> (Long pause, then H.M. chuckles) "Well there... (chuckles again)... there was one fella that he and his family had moved from their house into where I lived, in Hartford, and his name was McCarthy. And his first name was Charles. (chuckles again at the idea that a friend of his could have the same name as ventriloquist Edgar Bergen's dummy)
> "There was another fella of course who drove a car that was older than I was and he lived on Essex Street. And we went to the same school, we graduated together. St. Peter's School in Hartford. We graduated together and,

well, we'd go up to Springfield and just tour around in a way. Charlie would be with us...and he would..."

"What was his name?"

"Hamilton. Harry Hamilton."

"Did you ever have a girlfriend?"

"Well...uh yes. It was Margaret."

"Tell me about Margaret."

"Well actually, she had a nickname. Polly. She was ahead of me in school. She was in second-year high when I was just going into my first year. She was a year ahead of me in school. But I first knew her when she used to live on Franklin Avenue in Hartford."

The conversation went on in this vein with H.M. telling Dr. Corkin how he once went roller skating in Miami, Florida, and how he used to go out to the farm as a child and shoot at targets or tin cans with a variety of handguns. Although soft-spoken and sometimes hesitant, H.M. could be any sixty-seven-year-old reminiscing about his childhood. His memories are not detailed, perhaps because he is unable to update them, but had you just met him, it's unlikely you'd suspect anything was wrong. But a different picture emerges when the conversation moves to the present, as the one with Suzanne Corkin did:

"When you're not at MIT what do you do during a typical day?"

"Uhh see that's...I don't remember things."

"Do you remember what you did yesterday?"

"No I don't."

"How about this morning?"

"I don't even remember that."

"Could you tell me what you had for lunch today?"

"I don't know...to tell you the truth."

"What do you think you'll do tomorrow?"

"Whatever's beneficial."

"Have we ever met before, you and I?"

"Yes, I think we have."
"Where?"
"Well... in high school."
"In high school!"
"Yes."
"What high school?"
"In East Hartford."
"And what year was that about?"
"1945."
"Have we ever met any other place besides high school?"
(Long pause) "To tell you the truth I can't... no, I don't think so."

H.M.'s claim that high school is the only place he and Suzanne Corkin have ever met is of course backwards: they are not the same age and did not attend the same high school, but they have met hundreds of times at MIT over the last three decades. H.M. likely has an elusive memory of Dr. Corkin and places it back in a time from which he can retrieve memories. After all, he couldn't believe he met her in the sixties—he *has* no memory of the sixties.

"Are you happy?"
"Yes... well the way I figure it is, what they found out about me helps them to help other people. And that's more important. That's what I thought of being too: a doctor. A brain surgeon."
"Tell me about that."
"And as soon as I started to wear glasses in a way I said no to myself. Because [in] brain surgery you have bright lights and there would be an extra glare from the rim of your glasses that would go into your eye right at the moment you want to make... the most important severance in a way. You might just move a little too far and that person could be paralyzed on one side. So that's

why I didn't…"

"Are you ever sad?"

"No, I figure what is done about me, well…helps the doctors and the nurses and everybody—helps other people. I figure that's more important in a way. Because what they learn about me they can also do on someone else or *not* do on someone else."

"Exactly."

"Just a thought I had…I wanted to be a surgeon myself. But as soon as I started wearing glasses I said no to myself. I said suppose an attendant were to mop your brow, hit your glasses and knock them aside. You could make the wrong movement then. And that person could be dead or paralyzed."

The last minute or so of that conversation portrays H.M.'s world more vividly than any other. He tells two versions of his decision to abandon the dream of becoming a brain surgeon, a story that is probably a kind of personal myth about his own operation; the time between those stories on the original tape is no more than thirty-five seconds. No sooner has the original version faded from his immediate memory than he tells it again, slightly differently. This is nothing new to the people who know H.M. well. Compare those two stories about becoming a doctor to this version, described by neuropsychologist Jenni Ogden in 1991:

H.M.: …At one time that's what I wanted to be, a brain surgeon.

J.O.: Really. A brain surgeon?

H.M.: And I said 'no' to myself, before I had any kind of epilepsy.

J.O.: Did you. Why is that?

H.M.: Because I wore glasses. I said, suppose you are making an incision in someone (pause) and you could get blood on your glasses, or an attendant could be

> mopping your brow and go too low and move your glasses over. You could make the wrong movement then...

In this particular conversation, just as in the tape I've been quoting from, he returned moments later to retell the story, this time without reference to the blood on the glasses. This is H.M.'s world, a set of oft-told tales, varying almost not at all from one telling to the next, incorporating (according to those researchers who work with him) the same facial expressions and gestures.

H.M.'s childhood memories, as limited and possibly diminishing as they are, present a wealth of information compared to the period since his operation in 1953. It's not exactly true to say he remembers absolutely nothing that has happened since then: he is dimly aware of the assassination of President Kennedy, although at different times he has called him "Robert," and in one instance, although he was able to identify Elvis Presley as a recording star (Elvis's first record came out after H.M.'s operation), he thought that perhaps Elvis had been killed by a bullet intended for the president, a twist on JFK's death that even the conspiracists have overlooked. But beyond these few mental bits and pieces, H.M.'s current life is a blank.

H.M. holds the record as the most written-about person in medical history for good reasons, the most important of which is that until his tragic case, the role of the hippocampus in consolidating new memories was simply unknown. One neuroscientist told me that had H.M. not come along he and his colleagues would still be removing the hippocampi from rats and wondering if the amnesia those rodents developed was of any relevance to humans. H.M. established in black and white the connection between the hippocampus and memory, and the fact that he has retained a few scraps of memory since his operation might be attributable to the fact that there are still very small fragments of that organ left in place in his brain. Work is under way to image H.M.'s brain and establish once and for all whether or not the eight centimetres of hippocampus William Scoville reported removing was accurately measured and reported.

The hippocampus is now the target for a rich variety of scientific investigations—it even has its own scientific journal of the same name. And it is not just the human hippocampus that attracts scientific attention. Birds and animals that require good memories, especially for places, have been found to rely on their hippocampi. The black-capped chickadee is a good example. In a typical day these little birds might conceal a couple of hundred seeds inside pine cones, holes in tree trunks or even in rolled-up dry leaves, meaning that within their home ranges there are thousands of such hiding places. In a long series of experiments Dr. David Sherry of the University of Western Ontario and his colleagues have demonstrated that there is little doubt that chickadees remember where they have stored seeds. It's now known that chickadees and other birds who store seeds have a bigger hippocampus than expected for the size of their brain. Chickadees whose hippocampi have been damaged wander about relatively aimlessly, sometimes finding seeds, sometimes not. There is now growing evidence that any animal or bird with a big hippocampus is likely to have some special need to remember a lot of different places. These don't have to be food storage sites either: for male meadow voles (better known as meadow mice) a big hippocampus allows them to remember where exactly they might find each of their several mates. Their hippocampus is much bigger than that of the female. However, in the closely related—but monogamous—pine and prairie voles the female and male hippocampus is the same size. But even the meadow vole pales beside the male Merriam's kangaroo rat, which both mates with several females *and* maintains a number of different food caches. He spends his life desperately processing spatial information and has the hippocampus to prove it.

H.M. has also revealed that memory comes in different forms, and astonishingly, while his amnesia for facts, events and people is nearly complete, he is able to learn new things. This was discovered by Dr. Brenda Milner of the Montreal Neurological Institute in the late 1950s. She was one of the first brain experts to work with H.M., and on her second visit with him she brought a piece of equipment from the McGill University psychology lab, a mirror-drawing test consisting of two star shapes, one slightly smaller than the other, fitted together so as to leave a narrow gap between them. The test is to take a pencil and run it along the gap, but it is much trickier than it sounds because you are only allowed to watch what you are doing in a mirror. The natural tendency when you arrive at a turn is to move your hand—as seen through the mirror—the wrong way, but after several attempts most people learn how to do it. The funny thing was that H.M. was able to master the mirror test too. After thirty trials over three days he was accurately tracing the star outline without much effort. But he had

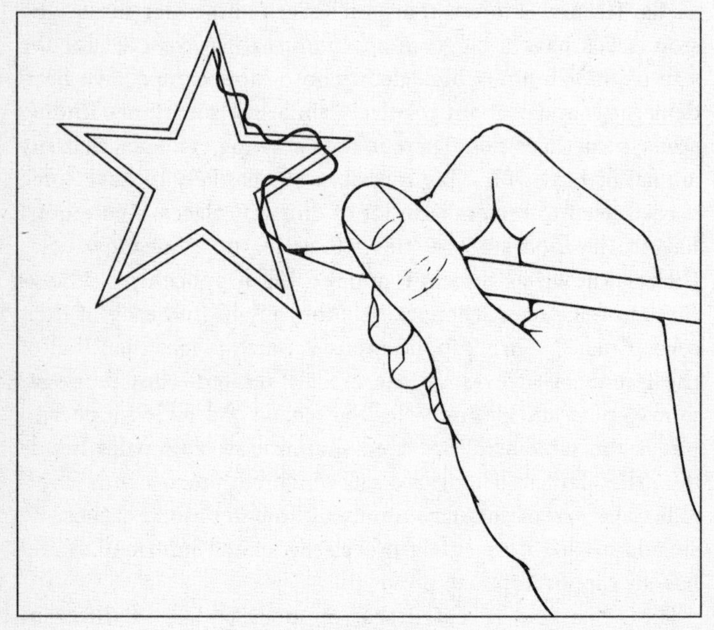

The Mirror-Drawing Test

no idea that he had ever done the test before and felt completely unfamiliar with it (as he undoubtedly did with Dr. Milner). H.M. has learned other skills over the years and his ability to do so has made it clear that there are different kinds of memory. The memory for learning a skill is separate from the learning of new information, and in his case is unimpaired. It is the difference between remembering how to ride a bicycle and remembering the colour of your first bicycle.

H.M.'s story sheds another light on memory that while difficult to quantify might be the most important lesson of this whole story. We think of memory as being the ability to re-create moments from the past, like H.M.'s memory of driving to Springfield with his friends. But memory is much more than that: it is integrated into every thought you have, at every moment. As you read these words, you are aware of where you are, what day it is, what you were doing in the hours before beginning to read, and can probably list a series of things you have to do once you put the book down—memory permeates your thoughts. H.M. has none of those things in his mind. Loss of memory doesn't seem to diminish intelligence, at least as measured by an IQ test: H.M.'s scores have actually risen since his operation, probably because he no longer suffers from the disruptive epileptic seizures. But without memory you cannot identify with a particular time and place; nor can you think about the future. Without it, you are not just unaware of what you had for lunch, you would be unaware of who you are. H.M. knows only who he was, and the portrait he retains is old and fading. Even at that, H.M.'s life is not impossibly remote from yours and mine. Suzanne Corkin, the scientist who has worked with H.M. most intensively, tells how she accompanied H.M. to his thirty-fifth high school reunion in 1982. A number of his classmates were delighted to see him, but he recognized none of them, either by name or face. This might at first seem surprising, because H.M.'s operation came long after he'd graduated from high school, and so memories from that time could have been intact. Either his amnesia has extended back into the pre-surgery period, or perhaps the seizures he was suffering during his high-school years, together with the high doses of anti-convulsant medication, interfered with the establishment of memories in the first place. In any case, H.M. wasn't the only attendee who had memory problems. A woman at the reunion claimed she didn't know anyone in the room either, and in her case it appeared to be simply a function of age.

Was H.M.'s surgeon, Dr. William Beecher Scoville, the villain in this piece? He has never, to my knowledge, been accused of any wrongdoing. He acknowledged from the beginning that the surgery was "frankly experimental," and most experts would likely have agreed that desperate measures were justified, given the intractability of H.M.'s seizures. It would be anachronistic to criticize the operation on the grounds that such invasive brain surgery seems a devastingly blunt instrument—in any era some of the medicine of forty years before will seem primitive or even wrong-headed. When Scoville realized the catastrophic effects removal of most of the hippocampus had had on H.M., he campaigned actively in the medical community against its use, referring to the "grave danger" posed by this kind of surgery.

But while there appears to be no blame to be laid specifically in H.M.'s case, there is evidence to suggest that memory loss as a side effect of this kind of surgery could have been noticed prior to H.M. If it had, H.M.'s story might have been quite different. Some background: in the first few years of the 1950s William Scoville performed many removals of the hippocampus and its surrounding tissue in an attempt to ameliorate the worst symptoms of schizophrenia. In this sense, removing the hippocampus and part of the temporal lobe was just another kind of lobotomy. Scoville practised two versions of this surgery: one that could be called moderate, in which about five centimetres of brain tissue were removed, and the second, more radical kind (that H.M. had) where eight centimetres were separated and suctioned out. The difference is a couple of centimetres of brain on both left and right.

H.M. was the only patient who had the surgery to eliminate the seizures of epilepsy, and he had the radical version. It was immediately apparent that he had suffered the "very grave recent memory loss" that Scoville later announced to his medical colleagues. H.M. couldn't even find his way back to his hospital room. But he wasn't the first person to have had this radical removal of the hippocampus. In December 1952, several months before H.M.'s surgery, Scoville performed the identical operation on a fifty-five-year-old woman with manic depression (now called bipolar affective disorder). She was a clerical worker known as M.B. who was confined to the Connecticut State Hospital because of her severe mental disability. She too suffered an almost complete loss of recent memory as a result of the operation, but this fact was not discovered until she was given a psychological test nearly a year after her operation—and two months after H.M.'s. Scoville himself later attested that her memory loss had escaped notice at first "because of her disturbed emotional state." She must have been seriously disturbed, because it didn't take long for testing (when it was finally given her) to reveal that she was another H.M. When Dr. Brenda Milner tested her a second time in 1955 (M.B. was one of ten of Scoville's patients Milner was following up) she couldn't even remember coming to the examination room from another building. Could such a memory loss have been masked by disturbed behaviour alone, or was memory loss not something that the medical community was expecting, or looking for, in these surgical patients?

There is evidence that memory loss was not exactly uppermost in the minds of those evaluating these novel and (at least at the time) promising treatments. While M.B. was the only patient of William Scoville's who had suffered dramatic memory loss prior to H.M., she wasn't the only patient whose memory had been affected to some degree by removal of part of the hippocampus. Among those patients who had had the more moderate removal of five centimetres of hippocampus on both sides were several with impaired memory, some of them serious.

Brenda Milner found one woman with an IQ of 123 who was able to name the year, but not the month or the day. But two such patients stand out because they were evaluated twice: once for the effect of the operation on their psychosis, and again later by Brenda Milner for their memory. Both were found by Milner to have suffered significant memory loss, and while she correctly describes it as being less severe than H.M.'s, it is still a degree of disability that left one of these patients unable to name her employer or to recall in the afternoon a conversation she had had in the morning. But memory simply isn't mentioned in the earlier psychiatric post-operative evaluations. Instead, Scoville and his colleagues rated any changes in the patients' psychosis on a numerical scale from minus 1 (worse after surgery) to plus 4 (markedly improved and discharged to home). One of these women whose memory had been damaged rated a plus 1; the other, a plus 4. Both had been operated on in November 1951, nearly two full years before H.M. Unfortunately Brenda Milner's memory evaluations weren't conducted until 1955.

So there were three patients who had suffered memory loss from moderate or radical removal of the hippocampus before H.M., but those memory impairments were not revealed until after his case alerted everyone to the severity of the problem. Yet even then, the surgery didn't stop. On May 13, 1954, eight months after H.M.'s operation, William Scoville was called to Chicago to operate on a schizophrenic doctor. He performed the moderate version of hippocampal removal—about five-and-a-half centimetres—and the next day his patient could recall myriad details of his medical training but didn't know where he was or why. He never regained his recent memory. It was shortly after that that William Scoville began his successful campaign for an end to surgery which removed pieces from both sides of the hippocampus. Why did Scoville turn once again to the operation when he already knew about H.M. and his female predecessor, M.B.? In his defence both of those patients had had the radical version of the operation, while this was the more moderate removal. But even to consider removing five centimetres of hippocampus before any detailed psychological testing had been done on the previous patients who had had that surgery requires at the very least an unshakeable confidence—or faith—that nothing would go wrong.

There is a necessary postscript to the story of H.M., which is that, however wrong his operation went, surgery to relieve epilepsy is alive and well and in most cases deservedly so. Precise brain mapping followed by removal of a piece of brain tissue that is triggering seizures can turn the life of an epileptic patient around. Losing a little piece of your temporal lobe can be a small price to pay for relief from incapacitating seizures. But as always, the first missteps towards techniques like this are the most painful, as they involve real patients. Never was that more true than for H.M.

16
Here and Now

H.M. has been terribly incapacitated by his inability to commit anything to long-term memory, but his short-term memory is intact, and that is all that keeps his life from being completely hopeless. The difference between the two kinds of memory is more than just how long each remains in the brain—they are managed by two different brain systems, and it is the anatomical distinction between them that allows H.M. to retain his short-term memory. He may not have a hippocampus, but he has whatever it is that deals with short-term memory. There is an oft-told story about H.M. that illustrates the contrast between the two. Dr. Brenda Milner was working with H.M. one day when she gave him a number to remember—584—then left the room for several minutes. That is a time span far beyond the limits of short-term memory, but when Dr. Milner returned, H.M. was able to repeat the number immediately. Although this might appear to have been some miraculous return of his long-term memory, it was not; in her absence he had had nothing to distract him and so was able to rehearse 584 over and over, refreshing it on his mental screen before it could fade. As soon as she began asking him new questions, the number vanished from his mind. But as long as there was an opportunity for him to rehearse the number, he was as good as anybody would have been at remembering it.

If short-term memory were dependent on the hippocampus and the parts of the brain immediately surrounding it, it's a good bet H.M. wouldn't have been able to remember the number 584 for five seconds, let alone five minutes. Short-term memory must be located in some part of the brain that is intact in H.M. Or at least nearly intact: the number of digits he can keep in his mind dropped from six to five a few years ago, a change labelled by those studying him as a normal age-related decline, an acknowledgement of the fact that H.M. is now well into his sixties. But other than this slight slippage in capacity, H.M.'s short-term memory is operating normally, small consolation indeed for a man whose life has become one in which "every day is alone in itself, whatever enjoyment I've had, and whatever sorrow I've had."

You might think that H.M.'s life would have been even more difficult if he had suffered the other side of the memory loss coin: loss of short-term memory. In that case it would seem that an intact hippocampus might sit forever waiting for new images and sensations that would never arrive because they would never be captured by the disabled short-term memory system. You might think that, but oddly enough there have been cases reported of severe short-term memory deficits—for example the inability to remember any more than two numbers at a time—and for the most part the people with them seem to have no trouble creating long-term memories. These patients have therefore sparked controversy over whether anything resident in long-term memory must pass through short-term memory first to get there.

Short-term memory is a fleeting memory of limited capacity, very good for things like telephone numbers. It is perfectly adequate for the task of looking up a number and keeping it in mind as you dial. In fact if you were left completely undisturbed, the way H.M. was when he was thinking about 584, you could rehearse a telephone number for several minutes and keep it alive in short-term memory. You could also do what H.M. can't—commit the number to long-term memory, at which point you could stop rehearsing it. A telephone number is an appropriate piece of information for short-term memory, because seven digits is close to the maximum number that we can "keep in mind" at any one time. The idea of seven as a magic memory number was first broached in the 1950s by psychologist George Miller. He actually settled on seven plus or minus two to allow for the considerable variation seen from person to person, and he also pointed out that although the limit of seven might well apply to digits or letters, it's safer to call these individual pieces of information "chunks." A chunk is something that is remembered in one piece, and while it can be a single number, it could also be three (if the first three digits of the telephone number represent a well-known exchange), four (a famous date) or even seven (when a clever pizza company transforms a seven-digit number into a single unforgettable chunk.)

It would be misleading to suggest that short-term memory evolved for remembering telephone numbers and little else. Psychologists like Dr. Meredyth Daneman at the University of Toronto have demonstrated the importance of short-term memory for reading, where the ability to remember words for a second or two is the only thing that makes comprehension of a long and complex sentence possible. I will describe her research in more detail later.

How short is short-term? This question isn't as simple as it sounds, because all of us, like H.M., can use rehearsal to hold chunks of information in our minds for very long periods of time, much longer than the short-term memory store theoretically allows. If you must, you can keep the telephone number in your mind while you move from room to room looking for the nearest telephone, but in doing so you are replaying the number again and again in your mind until the number is dialled, at which point you let it go. It's likely that most of what is casually labelled as short-term memory involves rehearsal, and that short-term memory without rehearsal, a rare bird indeed, is actually much more transient. Dr. Paul Muter at the University of Toronto designed an experiment in the late 1970s to determine exactly how long short-term memory lasts if there is no opportunity to rehearse, and he wasn't kidding when he called the report he published, "Very Rapid Forgetting."

In Dr. Muter's experiment, volunteers sitting in front of a computer saw more than a hundred events appear on the screen, each of which began with a set of three random letters displayed for a second. Most of the time the letters were followed by a two-second pause, then the word "LETTERS?". They were then to repeat the set of three they had seen a couple of seconds earlier. At other times, the three letters were immediately followed by a set of three random numbers, and the experimental subjects were then required to count backwards by threes from that number. By doing this, they would be unable to rehearse the letters. Neither of these versions of the experiment actually led to an estimate of how fast we forget. Only a few very rare "critical" trials did: these were a combination of the two above in that three letters were flashed, then the three numbers, and then the word "LETTERS?". So participants first registered the letters in their short-term memory, then concentrated on counting backwards, only to be asked practically immediately to remember what the letters were. And they didn't remember very well. As little as two seconds of backwards counting virtually wiped out the memory of the letters. The everyday version of this experiment would be looking up a telephone number and then immediately begin a conversation with someone. Could you remember the number?

Many more demonstrations of the fleeting nature of short-term memory have come from reading. In one study people were stopped in the middle of reading and asked to remember the order of words in a phrase from the previous sentence. They were much less likely to get it right than were people who were reading a slightly altered text which took that same phrase from the previous sentence and placed it at the beginning of the sentence they were still reading when stopped. The meaning of a phrase may not become clear until the end of the sentence, so presumably it must be remembered until that time. The time of forgetting must have been no more than a couple of seconds in this case as well. Two seconds seems less like memory than some sort of fleeting after-image, but this is only because words like "memory" and "forgetting" are normally used in reference to events from past days or years. Even to hold an image or idea or phrase in your mind for a couple of seconds' worth of conversation requires something beyond just taking it in, and that something is memory. We should not be surprised that short-term memory is measured in seconds. Common sense would tell you that it must be extremely brief: the amount of information flooding our senses and registering in the brain, however momentarily, is staggering. Ninety-nine per cent of it is unimportant and should be forgotten as quickly as possible.

So is short-term memory some shallow receptacle in the brain that holds only so much information for only so long—"so long" being a mere second or two if there is no opportunity for rehearsal? Far from it. Much has been revealed of this memory system over the last couple of decades, and its complexity and power have persuaded researchers to change its name to "working" memory, and more important, to divide it into three or more different subsystems. Each group of researchers designates or defines them in different ways, but one popular model of working memory divides it into three impressively titled parts: the "central executive" and its "slave" systems, the "phonological loop" and the "visuospatial scratchpad," names that put most other forms of techno-speak to shame. Each of the two slaves is specialized to handle its own particular kind of information; the central executive pays attention to their work and relates it to information from other parts of the brain.

The phonological loop is both a storage place and the part of working memory reserved for repeating the telephone number over and over to yourself using inner speech, the little voice that you hear in your mind's ear right now as you're reading these words. That it is auditory in nature is suggested by how much more difficult it is to remember a list of words which sound the same, like "man," "cap," "can," "map," "mad," than it is to remember a distinctive set of five words, like "pit," "day," "cow," "pen," "rig." Your recall of lists like this is also greatly impaired if you are listening to extraneous words at the same time, even if those words are in a foreign language. Funnily enough, bursts of noise have no effect on remembering, showing that mere distraction isn't enough. Alan Baddeley, a psychologist at the Applied Psychology Unit at Cambridge University and one of the foremost proponents of this three-part working memory system, argues that no matter which language they are from, words force their way into the phonological loop, disarranging the information that is already there and causing some of it to be forgotten.

The ability to maintain the maximum number of words, numbers or letters in working memory is dependent to a surprising degree on inner speech. As I mentioned earlier, most people can hold in their working memory seven (plus or minus two) chunks of information. Alan Baddeley has established that this number is equivalent to the number of chunks a person can *say* in two seconds, so strong is the link between retention in working memory and repetition of the words to yourself. Does that mean that if the language you use for speaking and mental rehearsal has very long words for numbers that you'll be handicapped when it comes to remembering as many as possible? Apparently it does. An experiment published in 1980 showed that people who are bilingual English-Welsh can hold fewer numbers in their working memory when using Welsh than they can in English. The differences in the words don't seem on the surface to be that great (the English one, two, three, four and five translate to *un, dau, tri, pedwar, pump*) but the people in this experiment, even though they considered themselves to be more competent in Welsh, took longer just to read numbers in Welsh than English. And the difference in the number of digits they could remember, having heard them at the rate of one per second, was significant: 6.55 in English versus 5.77 in Welsh. This experiment underlines the auditory nature of the memory for numbers: even if they are presented one by one on a screen, they are remembered by the sound of their name.

Hearing irrelevant words while you're trying to keep something in mind illustrates that the phonological loop portion of working memory depends on repeating the information over and over to yourself. In contrast, interfering with the phonological loop has no effect at all on chess players' ability to remember the positions of the pieces on the board, showing that in this case the slave of significance is the visuospatial scratchpad. It performs the same role for visual information as the phonological loop does for words: maintaining an important image long enough for something to be done about it. If the scratchpad is diverted by having to track a moving target on a screen or read (not listen to) irrelevant material, the memory for chess positions is greatly impaired. It's worth pointing out that the example of remembering chess positions illustrates that none of these memory systems works in isolation—each depends on additional information a person can bring to the situation. A chess grand master remembers the positions of chess pieces on a board much more accurately than a novice, but only if the placement of the pieces represents possible mid-game positions. If the pieces are distributed randomly, grand masters are no better than anyone else. Obviously their accurate memory of realistic positions relies on a working memory of normal power which can tap into a vast resource of experience, not an extraordinary working memory in and of itself. This suggests that so-called "photographic" memories might have much more to do with long-term than working memory. Those unusual people who can remember every detail of a scene they saw yesterday are not employing their working memory to do it.

The more complex a test of memory for chess positions is, the more likely it is that the third element of working memory, the central executive, is called into play. The central executive sits above the slave systems (metaphorically), surveys the material in them and determines what if anything will have attention directed to it. It exerts choice where the slave systems merely store and rehearse. If you are in the act of repeating the telephone number to yourself and someone asks you to say the fourth number out loud, or to add the seven digits together, the central executive steps in. If players are asked to choose the best next move on a chessboard they have just seen, and at the same time are required to do a second attention-diverting task like uttering random numbers or letters, their central executive will be overloaded. Not only will they have trouble with the choice of a next move, but sequences like 3,4,5,6 will start to appear in their "random" numbers.

Meredyth Daneman has focused her research almost exclusively on the central executive, and especially the kind of problem exemplified by the talking chess player: overload. Through her research on reading she has painted a picture of the central executive as a place where storing and processing information goes on side by side, and even depends on the same resources, so that too much of one can interfere with the other. In a sense this view of the central executive likens it to an early version of today's personal computers, in which it was impossible to store much information and do complicated manipulations of that information at the same time. If you are pre-computer, then think of the central executive as a single sheet of paper on which you must do a series of relatively difficult calculations. As long as you know short cuts, there will always be plenty of space left on the paper to write the answers. If you don't, you run out of space.

Nearly fifteen years ago Daneman and colleague Patricia Carpenter devised a method for illustrating how crucial the central executive is to reading. Reading makes more demands on memory than you might realize, because at any moment you can see only five or six letters beyond the one your eye is fixed on. This suggests that we read word by word (the claims of speed readers notwithstanding) and therefore rely heavily on our memory of what has gone before to have any hope of understanding what we are reading. Given that working memory seems to have limited space available, reading theorists had argued for some time that the difference between good and poor readers would simply be a difference in their working-memory capacity. Yet no tests of such capacity correlated very well with actual reading performance. Some people who could only remember three digits instead of seven were perfectly good readers and vice-versa. Daneman and Carpenter took the idea of working memory as a processor *and* storer of information and argued that it's not a question of how much space there is, but of how much space is left over after the information processing has taken place. They tested this idea by having subjects read aloud a series of sentences (printed individually on cards) and then try to remember the last word of every sentence. If the sentences chosen were from this paragraph, the final words so far would be "reading," "on," "reading," "capacity," "performance," "vice-versa," "place," and "sentence." Most people can remember no more than five such words, and the majority can manage only two or three. What's interesting is that the people who best remember these sentence-ending words are also those who read faster and score higher on tests of reading comprehension. Meredyth Daneman argues that in this test people are doubly taxing their central executive by both analyzing the sentences and remembering the words, and that a poor performance means that an inefficient central executive (as so many executives are) takes up so much space for processing that there's not enough left for storage.

On the other hand, those people with an effective central executive have plenty of storage space, and are better not only at remembering the final words of sentences—a task that is seldom very useful—but also at grasping with ease the meaning of a tricky sentence like "Although he spoke softly, yesterday's speaker could hear the little boy's question." Too little storage in the central executive and a reader would have to keep going back over that sentence to get it. In another experiment, readers were asked to remember a string of numbers while they read (and tried to comprehend) sentences like "The cat that the dog bit ate the mouse." When reading the sentence by itself, most people's eyes pause at the difficult point: right between "bit" and "ate," but their eyes remained there longer and longer as the string of numbers they were keeping in mind grew. But again there were differences between subjects: people who were the best readers could remember the most numbers at the same time as they read the sentence, but the worst readers had their hands full with the sentence itself, all of which suggests that the better your central executive, the more space it will leave untouched for storage, and the better reader you will be.

Loss of the central executive is apparently one of the many distressing symptoms of Alzheimer's disease. Patients with this dementing condition can be capable of remembering either sets of words or the position of a target on a screen (each of which would presumably be the job of one of the slave systems) but will then perform very poorly when asked to do both at the same time, something their central executive should be doing. As the dementia progresses, they may maintain their ability to do the individual tasks but fall farther and farther behind when doing the combination.

One of the mysteries of working memory has been its whereabouts in the brain. Whereas H.M. and other amnesiacs have proven that the hippocampus is essential for creating new permanent memories, it apparently has no role to play in working memory. Experiments where monkeys see a momentary image on a screen, then several seconds later are required to look directly at where it was, have shown that the frontal lobes play a major role in the ability to do this. There are even individual brain cells that play a role in each of these steps: some fire when the image is on the screen, some fire to direct the eyes to that place after the image has disappeared, and most interestingly some fire in between, when the screen is blank. Those are apparently the cells that are keeping the location in mind. In the summer of 1993 a group of American researchers extended the analysis of working memory to humans by publishing images of the brain's visuospatial scratchpad at work. The people taking part in this study were shown three dots on a screen for a fifth of a second, then were forced to wait three seconds before a circle appeared. They had to judge whether the circle would have encompassed one of the dots or not. As they were doing so, a PET scanner was recording those places in their brains where blood flow had suddenly increased. There were four main areas hard at work, all of them on the right side of the brain. One was in the frontal lobes, just as would have been expected from the monkey experiments. There was a second just behind that, a third towards the back right of the brain in the parietal cortex, an area known to be involved in locating objects in space, and a fourth in the visual part of the brain at the back, an area known to be involved in the perception of objects. What is unresolved is how these relate to each other: are the frontal lobes retaining the memory of the image, on the basis of information coming from the other brain areas, or are all four acting simultaneously to create an inner picture? A separate study has painted a picture of the phonological loop in action. People in this case were asked to keep six words in mind; again several different places in the brain were active, but these were different from those engaged in the scratchpad, although once again the frontal lobes were prominent.

Every parent knows that children around eight months of age or so develop the understanding of the permanence of objects: put a ball in one of two boxes while a child watches, then distract the child for a moment, then ask the child to find the ball. Very young children have no idea, but by eight months the child is able to remember which box to choose. At that time in the child's life the frontal lobes are just developing adult nerve circuits, suggesting that every human needs fully developed frontal lobe circuits to have an operating working memory, and prior to eight months of age they just aren't there.

Although the traditional view of the capacity of working memory (or at least the phonological loop) is that it is limited to seven chunks of information, give or take a couple, it is possible to extend that with training, sometimes dramatically. The classic case was that of a young undergraduate student at the Carnegie-Mellon University of Pittsburgh in the late 1970s. He is known only by his initials, S.F.—he died of leukemia shortly after these tests were done. When he began training his memory, he was completely unexceptional, being able to remember no more than seven numbers at a time. He practised remembering numbers three to five hours a week for twenty months, and in that time his memory span for digits went from seven to an incredible eighty! That means that if he were read a hundred numbers in a row, one a second, he'd usually be able to remember the first eighty. Prior to S.F., the record among memory experts—people who were recorded in the memory literature precisely because they had phenomenal memories—was eighteen.

The curious thing about S.F. was that he was given no coaching whatsoever, but came up with memory strategies himself. He happened to be a good long-distance runner, and already by his second week of training he had noticed that numbers sometimes came in groups that he would recognize as running times: 3492 became "3 minutes and 49.2 seconds, near world-record time for the mile." After he had run out of good running times, he began to group sets of numbers as ages and dates. Then he was able to group these sets of three or four numbers together as "super-groups," so a twenty-five-digit number became three sets of four digits plus three sets of three plus four at the end which he simply rehearsed. And there is absolutely no evidence that he was anything special. After his death the same group of researchers trained a second undergraduate in the same way, and he reached 106 digits. He likely progressed faster because he too was a long-distance runner, and he was taught S.F.'s trick of remembering numbers as running times when he began.

The problem with these examples of exceptional memories was that they were very narrow. Both S.F. and his successor D.D. were great at numbers but no better than the average person at words, because they had invented no system for them. And so their super-memories had little apparent application. But that is not the case for another instance of augmented short-term memory: the phenomenal ability of cocktail waitresses to remember drink orders. (In the particular experiment I'm about to discuss, conducted as it was in the early 1980s, all the waitpersons were female.)

Henry Bennett of the University of California at Davis designed an experiment to test the memories of students and waitresses for drink orders. To adapt the experiment to the psychology lab, he created a miniature cocktail lounge complete with tiny chairs and tables and dolls 9 1/2 centimetres tall (about 3 1/2 inches) as customers. The dolls were a variety of female and male, bearded or clean-shaven, bejewelled and not, thirty-three in all. The drinks were small rubber stoppers (used for laboratory glassware) with flags on top, each flag bearing the name of the drink. And finally, to complete this bizarre scene, there was a cassette tape with a series of orders for drinks, spaced two seconds apart: "Bring me a margarita...I'll have a Bud ..."

The amazing thing about this study is that when the investigators brought this whole miniature cocktail lounge—in a suitcase—to a bar and asked the waitresses there if they would be willing to take part in a memory experiment, they weren't laughed out of the place. When the waitresses gave their consent—only one refused to take part—the miniature lounge was laid out on a table, the miniature customers seated on their miniature chairs, and the tape turned on. How did the waitress know which customer was placing the order? "As each voice on the tape ordered a drink, we would wiggle one of the dolls indicating that doll was ordering the drink." This was done for seven, eleven and fifteen orders, placed either in order or at random around the tables. Roughly the same procedure was followed for undergraduate students, except that they played the game in the less-stimulating surroundings of the psychology lab.

Not surprisingly, waitresses remembered drink orders better overall than students: 90 per cent correct versus 77 per cent. The difference was especially marked with orders of eleven or fifteen drinks. As Dr. Bennett points out, this is an out-of-the-ordinary memory feat that is accomplished routinely in an ordinary job. In fact, the demands in the real cocktail lounge are much more difficult: the waitress must collect orders, rearrange them for the bartender, get more orders, pause for conversations, then pick up the drinks and take them to the tables. A good waitress does this all in her head and never has to perform an "auction": stand at the table and ask, "Who ordered the Bloody Mary?" Interviews with the waitresses revealed that they felt their memories to be more accurate the busier they were; as one said, "Then I'm in the flow." They also felt that tips were better if they remembered drink orders accurately, apparently because some customers (desperate to perceive attention where none exists) interpret accuracy as a form of flattery. But the most interesting revelation wasn't a revelation at all: most of the waitresses could not explain exactly how they were able to perform magnificent feats of memory. There was apparently a variety of strategies, including mental rehearsal of the order (using the phonological loop) and keeping track of the position of the guests at the table. But the most common strategy was to try to anticipate what each customer was going to order; by doing so the waitresses apparently focus their attention on the face and dress of the customer and establish a link with a drink that way. Three waitresses went so far as to testify that after a few months on the job, "customers start looking like drinks." Sounds like the visuospatial scratchpad to me.

The lesson from this study is probably that cocktail waitresses (and other memory experts) are able to stickhandle around the problem of the limited capacity of working memory by anticipating the future. They don't just rehearse drink orders as they come; they predict what these orders will be, and then match the actual orders to those expectations. If they are at all accurate in their predictions, the information they are required to remember is already in their minds before the customer speaks. In a sense they are turning what looks like a challenge to the capacity of their working memory into a task that draws on the much larger reserves of their expertise, their long-term memory. Regardless of the explanation, studies like this raise the startling possibility that there may be no upper limit to the number of different orders waitresses can remember. One waitress, who had to work a New Year's Eve alone when two others called in sick, remembers serving about 150 people that night: "By the end of the night I knew what every customer was drinking. I'd just stand by the bar, looking for hands and give the bartender the order. I really don't know how I did it."

17
Then and There

Even the best cocktail waitperson won't remember much, if anything, about the drinks that were ordered last week. A week is well beyond the limits of working memory, and any memory of events that long ago or longer is something completely different. For simplicity it can be called long-term memory, but as we will see, there is very little in the brain that can be viewed as a permanent store of accurate recollections of the past. We preserve only a minute fraction of all the incoming sensations and information to which we are exposed on a second-by-second basis, and even that sparse collection is whittled down as the days, weeks and years pass, and the memories that survive this process are almost always different—sometimes dramatically—from the original event. What is preserved in memory, what is lost, and why? When we know the answers to these questions we will be a lot closer to understanding what it means to be human.

The difference between working and long-term memory is more than a case of fleeting versus permanent, or discarded versus recorded. The two memories are handled by separate systems in the brain. Remember that H.M. has a perfectly adequate working memory: he can probably remember a telephone number for about the same length of time as you can. But his missing hippocampus prevents him from retaining any information for a longer period than that, except in the rare case where a person or an event is presented to him over and over again through decades. The hippocampus is where long-term memory begins. It is a relatively ancient part of the brain (unlike the highly expanded and folded cerebral cortex) and in all those living things which have one it plays some role in memory. In humans it has assumed even greater importance than in most other creatures: we apparently need it to establish any permanent or long-term memories. Without the hippocampus we would live in the eternal present; with it we can make the incredible journeys into the past that our personal memories represent. But having said that, there is still a great deal of mystery about just what happens in the hippocampus and subsequently in other parts of the brain that turn one recent event out of many into something that you'll remember a week or even forty years from now.

Two of the great unanswered questions about memory relate directly to the process of making long-term memories. Where are these memories to be found in the brain, and what actually happens inside the brain when a new memory is made? The answer to the first is "probably in the cerebral cortex." There in the bulges and folds there are enough brain cells—billions—each with enough connections to the others—tens of thousands—to maintain the number of permanent records we have in our memories. What is eerie about memory is that while the loss of a part of the brain, even an entire hemisphere, may ruin a person's ability to use language, or identify common objects, or follow the logical thread of an argument, there are few if any cases where the loss of or damage to a specific part of the human brain eliminated certain memories while leaving others untouched. The Harvard psychologist Karl Lashley once concluded facetiously, after trying fruitlessly to locate memories in animals' brains, that learning (which of course is totally dependent on memory) "just is not possible." The elusiveness of memory traces have led to unusual claims, for instance that memories exist in the brain in the form of holograms. The virtue of this idea, which was taken seriously for some time, is that, unlike a photographic negative, a damaged hologram loses some fineness of detail but retains the overall image, and so might be the analog of the damaged brain. But extending this analogy to memories preserved by brain cells has proven to be a long and difficult stretch.

It may even turn out that a single memory exists in many places at the same time. A popular idea currently is that the memory of an event exists in as many parts of the brain as there were sensory impressions triggered by the original event. Imagine that you have a vivid recollection of a childhood picnic, with a panorama of colour, movement, tastes, smells, laughter and shouting, all woven into one brief snapshot of your life. How and where in the brain does that memoir take place? Most brain scientists and philosophers today argue that there's no use imagining a tiny little theatre in your brain where it's all played out, because you are then forced to add a tiny viewer, and who would that be? Furthermore, do the front parts of the brain, where it has been assumed the different bits and pieces of memories flow together into a single recollection, have the storage capacity necessary for a lifetime of remembrances? These areas have fewer brain cells than those parts farther back that are thought to feed them the information. The alternative that is gaining favour these days was originally put forward by Antonio Damasio at the University of Iowa. He suggests that each remembered event is composed of many different pieces, and these pieces never actually meet in the brain. According to Damasio, the visual portions of that memory of the picnic remain in the visual cortex at the back of the brain, the smells in that part of the right frontal lobe known to be associated with the memory of odours and so on. If this were so, and you were prompted to recall that picnic, all these different parts of the brain would reactivate their part of the memory, and then somehow you would experience them all, simultaneously, as that memory. That's the tricky part, called the binding problem: how does the brain reactivate all the sensory fractions of the whole event to make it come alive again? The task may not be as difficult as it sounds: when you look up from this book at the scene around you, the different aspects of the visual image, the colours, shapes and movement are all processed in different places in the visual cortex at the back of your brain, and yet somehow the product of these separate processing facilities is recombined to give you—instantaneously—a complete image. Memories might result from a similar recombination.

The stimulus for the memory might involve only one of the original senses you used, but once awakened, it could alert the others to fill out the complete memory. When I was a kid I played with Meccano, the pieces of metal filled with holes that could be screwed together to build miniature cars or cranes or houses. There were also wind-up motors that came with the set, so that the crane would actually lift something, or the car could move across the floor. If I were just to hear one of those motors today, I would remember the colour, shape and feel of the motor, even the slipperiness of the coiled spring that powered it. I could tell you exactly the size and shape of the key I used to wind it, and all of these images are decades old. Antonio Damasio would argue that the trigger for these details was the sound, and that circuits of brain cells in the auditory areas of the brain would connect to what he calls "convergence zones," which in turn would activate the other parts of the brain which preserve the tactile and visual parts of that memory. It would be just as easy for the feel of the key or the sight of the spring to trigger the rest.

If it were true that fractions of each long-term memory trace exist in the different sensory areas of the brain, or in the associated areas that combine them, this might be at least a partial explanation for the resilience of memories. It would be hard to wipe one out completely. What should sometimes be seen, however, is that certain kinds of brain damage could eliminate one aspect of a memory, for instance the smell or the sound. It is true that damage to the part of the visual cortex that deals with colour can leave a patient not only incapable of seeing colours but also unable to remember what colours were like. The world they see today is black and white; so are their memories.

What happens in the brain when a new memory forms? Most scientists believe that permanent changes must occur at the synapses, the places where brain cells communicate with each other. These changes are likely physical alterations of the brain cells such that they develop the cellular equivalent of an "itchy trigger finger." Think of Pavlov's dogs: they were trained to salivate at the sound of a bell by associating that sound with the odour of food. Drooling when a bell rings doesn't have the emotional resonance of the memory of the sun glinting off your true love's hair, but it's quite likely that the neural underpinning is the same for both. Today the famous Pavlovian experiment could be recast in synaptic terms: before training, those cells in the dog's brain responding to the odour of food would be linked—through scores of intermediates—to brain cells controlling salivation. The introduction of a third element, the sound of the bell, would trigger brain cells in the dog's auditory cortex, and while these cells might routinely dump a few neurotransmitters in the direction of the odour/salivation cells, whatever faint response that resulted would at first be lost in the general neural hubbub responsible for slobbering over the doggie chow. But the persistence of these sound signals and their simultaneity with the odour would reinforce the tenuous chemical effects, and soon the brain cells in the salivation loop would be responding to the bell as well as the odour.

Some of the molecular details of that general picture have been filled in. Picture a single brain cell at the heart of these circuits whose main job is to react to the sound of the bell and pass that information on. At the same time, it is "listening" to many other brain cells, including those making up the main smell-and-salivate circuits. When the bell rings and the odour immediately follows, this cell receives two nearly simultaneous signals, one from the bell (more or less directly) and one from the smell (third- or fourth-hand). When two neural signals arrive in rapid succession like this again and again, the recipient cell is chemically primed. The timing is critical because some of the chemical changes are transitory, and unless bell and odour are virtually simultaneous, the first of these changes has faded before the second arrives. The changing chemistry means that this cell now exaggerates its normal response to the sound of the bell, releasing many more neurotransmitters than it did formerly, and making a much bigger impression on all the brain cells it is in communication with, including the network controlling salivation. It now has a central role in the brain circuitry controlling salivation.

Over the long term, that cell might go further in establishing the permanence of this new role by disassembling and rebuilding its internal skeleton so as to expose receptors that were previously hidden inside folds or creases. These receptors could make the cell even more sensitive to the sound of the bell. It might build entire new terminals for receiving or sending neurotransmitters. These are permanent changes to create new communications channels, but notice that no new brain cells had to be recruited. Pavlov's dogs' brains established a permanent memory of the unusual association between a bell and food simply by retrofitting existing brain cells.

Changing the connections brain cells make with each other seems the most sensible explanation for the creation of new memories. Even though there are billions of cells available, using tens or even hundreds for each new memory would inevitably fill the memory banks to overflowing. Better to tinker constantly with the thousands of synapses on each. Another virtue of preserving memories as patterns of communication among brain cells is that the cells are permanent, just as many memories are. Nothing else in the brain has that kind of longevity, although that didn't stop scientists in the 1960s from arguing that memories could be recorded in the form of individual molecules. This thinking set the stage for one of the oddest episodes in the history of memory research.

In 1962 James McConnell published the results of experiments which suggested that worms called planaria could acquire memories simply by eating their trained fellows. Planaria are tiny flatworms which live in freshwater streams and have the charming ability not only to survive being cut in half, but to turn that to their advantage by having each half develop into a whole new worm. Simple as they are, they are capable of learning, and learning is not possible without remembering. McConnell first showed that planaria could be taught (in the same way Pavlov's dogs were) to associate a light with an impending electric shock. The worms were considered to have learned the connection between the two when they contracted their bodies in response to the light alone. McConnell then showed that if those trained worms were ground up and fed to untrained worms, these novice worms would then learn the association between light and shock quicker than normal. The implication was that in eating their comrades they had somehow consumed a "memory molecule" that gave them a headstart in learning.

This was obviously an incredibly exciting discovery, although needless to say its application to human learning wasn't immediately clear. Unfortunately the planaria experiments looked much less remarkable when it was shown that it didn't matter much what kind of worm, trained or not, was the entrée. Even when worms ate others that had never seen the light or experienced the shock, or had simply been handled by the experimenters, they seemed to learn quicker. So while it was true that learning was quicker after eating worms, it was probably facilitated more by a good meal than by any memory molecule. Too bad it didn't work out—it would have served as the theme for a few good B movies. (As the star football player threatened with being dropped from the team because of his low marks says to the egghead student, "Hey—I hear you've already studied for the big exam next week...")

Almost all the research aimed at explaining exactly what goes on at the molecular level when a memory forms uses simple animals with simple brains, ranging from marine molluscs through fruit flies to rats. At the same time psychologists continue to flesh out the details of human memory in the hopes that the two approaches will eventually meet. The study of human memory relies heavily on the introspection of volunteers, the same kind of "thinking-about-memory" of which we are all capable.

One of the features of human memory that becomes obvious when you reflect on it is that there is no instantaneous change from short-term or working memory to long-term or permanent, no switch in the brain that abruptly determines that the phone number you just memorized is now part of the permanent record. It takes months or even years for memories to be consolidated, and as long as they are in the grey area of not-yet-permanent they are vulnerable to forgetting. Imagine going out to dinner with friends. The night of the event you have the telephone number and address of the restaurant in mind, as well as the significant current events in yours and your friends' lives. But it is not just the waitress who is going to forget the details of the evening. The next morning you may be able to remember what everyone ordered, the topics of conversation, and anything that was particularly funny or irritating, but a week later, as long as there has been no rehearsal, many of the details that were so fresh will have faded, become confused with each other or will have disappeared altogether. A month later you will be hard-pressed to remember anything of the conversation (unless something extraordinary was said), and a year or two later you might not even remember going to dinner, or if you do, it will be a token memory, a factual record that you did go out with these people, rather than the colourful living recollection you had the morning after.

Numerous surveys have established that forgetting continues

at a fairly steady rate over the months and years after events took place and made their first entrance into our memories. One study suggested that if you were to look back and choose a benchmark year six or seven years ago, you would find that your memory for the events of that year had declined by about 6 per cent annually to the present. And that's not all: during this time the nature of the memory itself has changed. Later events change the significance of earlier ones, making them more or less likely to be remembered. Similar events coalesce, leaving behind a single memory which is accurate in its parts but inaccurate in the whole. (Yes you caught a snake at the cottage, but Aunt Lou wasn't there with you—she visited the next summer.) Two phenomena in particular appear as you allow your mind to range back over the memories of your earlier life, and both have occupied the attention of memory researchers. They are "flashbulb" memories, and childhood amnesia.

Flashbulb memories are those incredibly vivid mental pictures you have of events like the Challenger disaster or the Kennedy assassination. Canadians would probably add Paul Henderson's series-winning goal in the 1972 Canada–Soviet Union hockey series. People who experienced these events can tell you where they were when they heard the news, who they were with, what their reaction was and any number of additional details that do not accompany run-of-the-mill memories from the same era. It is as if the sensations of the moment were preserved photographically. Each of us has personal flashbulb memories as well for events like births and deaths in the family. There have been suggestions that the memory mechanism for such events might be different from that which preserves the more mundane recollections, a claim that if true would be very helpful for understanding memory in general. The most exciting thing about flashbulb memories is the amazing detail preserved in them. If the fidelity of that recall could be understood, it might then be possible to record many more memories of that intensity, if indeed we would want to.

Two psychologists at Harvard, Roger Brown and James Kulik,

coined the term "flashbulb memory" in the first in-depth publication on the subject in 1977. They suggested that for an event to become a flashbulb memory it has to have both novelty and "consequentiality"—it has to mean something to the individual. Obviously births and deaths in the family have both. JFK's assassination and the Challenger disaster are extremely high in novelty, but less so in consequentiality, particularly for Canadians. Henderson's goal? What can you say about a moment that becomes part of a nation's mythology?

Brown and Kulik borrowed a little-known theory called "Now Print!" to give some biological substance to their claim that flashbulb memories are those with high levels of novelty and consequentiality. "Now Print!" had been suggested by psychologist R.B. Livingston, and it argued that the brain would, in a moment, recognize the novelty of an event, judge the significance for the individual, and if both were high, issue the "Now Print!" order. It would preserve in memory not only the event itself, but all other ongoing mental activity. The beauty of this explanation is in this last part, which would explain why different people remember minute and totally irrelevant details when describing the same event, like the Harvard professor who remembered the feel of the rubber tread on the steps he was climbing when he heard of JFK's shooting. I remember hearing the shout accompanying Henderson's goal as I was supervising a chemistry lab for extremely reluctant graphics arts management students at what is now Ryerson Polytechnical University in Toronto. (If I had been smarter, we all would have been watching.) Why should it be important that I remember the actual room, or where I was standing? Why should the brain flashbulb details that have nothing to do with the central event? Brown and Kulik claimed that the "Now Print!" mechanism must have evolved at least hundreds of thousands of years ago, at a time when flashbulb events weren't delivered to you indirectly by the media, but were something you experienced personally. In that case, circumstances that might seem irrelevant today (where you were standing, what you had been planning to do) were actually part of the event, and so worth remembering.

Since that time, a debate over the significance of flashbulb memories has developed. Even Brown and Kulik pointed out that they aren't quite as dramatic as you might think—for one thing, they are far from photographic in nature. Bring your favourite flashbulb memory to mind as fully as you can, then ask yourself these questions. What shoes did you have on? (Only a memory of streaking would include that detail). What was the weather like? Who did you have dinner with that night? What were you doing five minutes (or an hour) before the event? A few questions like this should persuade you that a flashbulb memory is nothing at all like a complete record of the moment, and may not be, at least in content, that much more complete than standard memories.

The suggestion that they might involve a special memory mechanism ("Now Print!") bothered some psychologists, if only for the reason that creating two kinds of memory where one might suffice only complicates the job of figuring out memory, period. Through the 1980s a variety of studies claimed that flashbulb memories were not nearly as detailed and as accurate as their possessors believed (one scientific paper was titled, "Flashbulb Memories: Special But Not So Special"). These studies, even though they provoked their own controversy, planted seeds of doubt about the nature and quality of flashbulb memories themselves, let alone the existence of a special mechanism for making them.

Then in 1991 serendipity provided Baylor University psychologist Charles Weaver with a unique test of flashbulb memories. He was trying to discover if it were possible to form a flashbulb-type memory of an ordinary event, and on January 16th of that year he asked his students to remember everything they possibly could of the next time they met a roommate or friend. To this end they were to fill out a questionnaire detailing the time of the meeting, where it happened, what the student was wearing and so on. However, that very night

President George Bush ordered the bombing of Iraq. Suddenly Weaver had a unique experiment on his hands: two events, one of which, the bombing, could legitimately qualify as a flashbulb memory, and another, the meeting, that students might turn into one. So that week he had them fill out a second questionnaire about the bombing, and then tested their memories for the two events again three months and twelve months later.

Weaver found that the accuracy of the memories for the two events eroded in a pattern familiar to memory researchers: a decline in accuracy over the first three months, but little change after that. More interesting was that there were no significant differences between the memory for the meeting and the bombing campaign. The students had created their own flashbulb memory for the meeting, something that would otherwise have been a trivial event. But there were clear differences when it came to confidence in the accuracy of the memories. The students invariably believed their memories of the Bush announcement and subsequent bombing were more accurate than their memories of meeting their room-mate. Weaver concluded that a "flash" is not needed to make a flashbulb memory—having the will to remember will suffice—and that the really outstanding feature of flashbulb memories is the degree of misplaced confidence people put in them. This experiment also casts doubt on the idea that flashbulb memories are made in some special way. If students can turn a mundane event into a memory with flashbulb qualities on request, it suggests that focusing attention on any event is enough. Why we believe our flashbulb memories to be more accurate than they really are is as yet unanswered, although psychologist Ulric Neisser at Emory University in Atlanta believes they are links between personal events and the events of history, are extremely important to us, and so we need to believe they are accurate.

If the intrigue of flashbulb memories has been whittled away by experiment, the same cannot be said of childhood amnesia.

It is, as they like to say these days, a "robust" phenomenon, and it remains baffling. Childhood amnesia is the inability to remember very much if anything that happened during the first few years of life. There have been varying estimates of exactly what period it covers, but most researchers would agree that we are all amnesic about the first three to four years of life. Because it is universal, we seldom realize just how remarkable it is that we don't remember anything about walking for the first time, switching from breast milk to solid food, learning to talk—all events of monumental importance that simply disappear from the record, if indeed they are recorded in the first place. In a report published in 1993 the same Ulric Neisser mentioned above showed that childhood amnesia is not uniform: memories for highly charged events can be recalled from earlier ages. Neisser asked college students to remember the earliest experiences they could, and then asked their parents to corroborate those reports. He found that hospitalization (for operations like a tonsillectomy) and the birth of a sibling were sometimes remembered from as early as age two, while other seemingly important events like moving homes or a death in the family could only be recovered from age three or older. But even though these results push the curtain of childhood amnesia back from age three or four to age two, at least for certain events, it remains true that nothing at all is remembered from before the second birthday. The first two years of life seem a complete blank, with one small caveat: in this study there were no students who had experienced a hospitalization when they were one year old, so although they forgot the birth of siblings in that year, there is still the possibility that hospitalization at that age might be remembered. Some psychologists have questioned the reliability of the data in this study, pointing out that in some cases what seems to be a true memory might actually be a guess (Where did you first see your baby sister?" "At the hospital.") or information gathered later from videos, slides and/or family stories. If accurate, these critiques would re-establish three years as the earliest date for reliable memories, still leaving the question, "Why forget them at all?" unanswered.*

There have been many suggestions as to why we remember next to nothing of our first couple of years of life. Sigmund Freud was the first to bring attention to this amnesia, claiming that we repress our memories of infancy because they are full of guilt-inducing sexual content. A century later, explanations focus more on the shortcomings of information processing in the infant brain. For instance, the key memory structure in the adult brain, the hippocampus, is not fully developed at birth, and so memories may simply not be laid down until it is ready. But as Neisser points out, this can't be the whole answer. One of the puzzles of childhood amnesia is that children have a perfectly good memory when they are living through this blank period. Two-and-a-half-year-olds remember things that happened six months earlier, so why can't they remember those same events years later? It might be that children before the age of two simply don't record their experiences in the same way adults do. If you quiz young children about some recent event, like visiting relatives in another city, they are apt to focus on routine events that wouldn't be particularly memorable: "We had breakfast, then we played, then we ate lunch." Some researchers argue that because such events aren't unique in any way, the entire visit is forgotten. It is possible that the child's brain doesn't register information in the same way an adult's does and so does not incorporate the set of events into long-term memory that one might expect. Any adult who has spent time with a young child knows that their world is differ-

* As far as I know, all the research on childhood amnesia has concentrated on the forgetting of events. But we also learn skills when we are very young. Some of them, like walking, are never abandoned, but there must be others which are learned then forgotten while still in childhood. My son, Max, is two years old now, and long ago learned what to do when someone approaches with a bib. He extends his arms and bends his neck, ready for the bib to be tied around his neck. If I approach him with a bib when he is ten years old, will he remember what to do? Is there childhood amnesia for skills as well as events?

ent: they take note of sights and sounds we have long since learned to ignore. They are absorbed by a car horn on a busy street, or the passing shadow of a cloud on a sunlit lawn. It seems more than reasonable to me that their brains are simply not registering the world the way we do. Ulric Neisser argues that this might explain why the birth of a sibling, which is given a great build-up and is described by parents as a highly significant event and so is given an adult frame of reference, is remembered from an earlier time than a death in the family, a much more mysterious event for young children. In the case of the birth, the child's memory is organized for him.

It might be that areas of the brain important for storing and retrieving memories, especially the frontal lobes, are reorganized with time, and any memories that predate that reorganization are lost, like card files in a warehouse that has just converted to computer. Then the explanation might be the following: nothing absorbed by the very immature brain, from birth to two years, survives the processes of brain maturation; only extraordinary events from the ages of three and four will be remembered as the child's brain begins to take on the appearance of the adult version while still retaining some of its earlier organizational immaturity, and memory is normal from the time the brain is fully developed. It is the price we pay for having such big brains. Humans have more post-partum brain growth and development than any comparable species because it would be impossible to squeeze a fetus with a fully grown brain through the birth canal. The evolutionary choice was either to widen the canal to an awkward degree, or arrange to have the brain grow after birth. Childhood amnesia might be the consequence.

18
Feats of Memory

My favourite bit of mythology about the brain is the old saying, "We only use ten per cent of our brains." This statement is true in the trivial sense that when you are sitting down, you are not using the brain cells that command your legs to run. But that is not what is intended: it is an expression of optimism, the idea that if we only bothered to exert ourselves, we could start filling up that other ninety per cent and become a whole lot smarter in the process. But it is a curious idea. What sense would there be in having a brain that only works at ten per cent capacity? That would make a 1500-gram human brain the equivalent of a 150-gram monkey brain, providing of course the monkey is using all *its* brain. Did hundreds of thousands of years of evolution take place simply to build a brain with more empty shelves? There is no logic in it. On the other hand, our brains always seem to have room for more. What if you changed your routine and spent tomorrow memorizing a poem? Where would the words, rhythms, emotional associations and memories connected with that poem be stored if not in some as yet unused storage space? What if starting tomorrow you took up line dancing, or playing squash? The rehearsed movements and the thoughts which accompany them must reside somewhere, and since the brain is a finite organ, then does it not follow that a fair percentage of it must remain unoccupied at all times in anticipation of the arrival of this sort of new information?

Part of the resolution of this paradox must be that new information can be superimposed on top of old by virtue of the way brain cells alter their connections and route new information through existing networks. Even the humble sea hare, a kind of shell-less marine snail possessing a mere twenty thousand neurons is perfectly capable of learning to respond to something novel in the same way Pavlov's dogs did. They do it by first altering the responses of some of their neurons by chemical means, then over the long term actually rebuilding some of their neurons to accommodate these new reactions. But they do not have neurons sitting dumbly waiting for scientists to touch their siphons or shock their tails. It must be the same in the human brain: you do not have unoccupied brain cells waiting for the next pop song melody to come along. By the same token it may not be the tragic waste of resources you have always thought it was to have brain cells still dedicated to remembering the words of "Yummy Yummy Yummy, I've Got Love In My Tummy." They can be used to memorize new songs.

There is of course a limit. Numerous studies have established that forgetting happens all the time over several different time scales, and in fact forgetting is necessary to be able to stay sane and make sense of the world. Your brain doesn't even pay attention to most of the sensory information that arrives every second, and it is "forgotten" literally before you have noticed it is there. The bulk of the information that passes through working memory, like telephone numbers, the exact amount you paid for yesterday's two-for-one pizza and who scored last night for the Calgary Flames is unceremoniously dumped. And even when information passes through (or around) these gates and makes it into long-term memory, the decay of forgetfulness eats away at it for weeks, months, even years. Just imagine the set of vivid memories you have now compared to the set you had when you were half as old as you are. I'd be willing to bet that there would be very little overlap.

You might wonder why I claim that forgetting is essential to mental well-being, and that's not surprising. Most of us forget how well we forget and so find it hard to conceive of what it would be like not to. But it is possible to envision the inability to forget. Nearly three decades ago the Russian neuropsychologist Aleksandr Luria reported the case of a fantastic mnemonist, a memory performance artist, named Shereshevsky. Shereshevsky used to stand in front of an audience, beside a blackboard full of numbers which he would study for a few moments, then turn his back and proceed to recite them flawlessly. He once memorized a long, detailed and nonsensical mathematical formula and was able to repeat it verbatim fifteen years later, without being warned he would be asked to do so. The problem came when he had to give more than one performance in an evening, because even when the backboard was erased, he had trouble forgetting the numbers from the first performance when he began the second!

He had to go to elaborate lengths to wipe the numbers out of his memory. At first he would envision covering the blackboard with an opaque film, which he then imagined tearing off the blackboard and crumpling up in his hand, with the numbers still on it. But even after doing that the numbers would sometimes pop up again when he returned to the blackboard. He then tried writing down what he wanted to forget, hoping against hope that his mind would release its hold on anything that was recorded on paper—perhaps it wouldn't need to remember anything that had been written. The problem was that he continued to see in his mind's eye what he had written. Shereshevsky finally conquered the problem in a sudden and unexpected way: he was about to give his fourth performance one April night and was dreading every minute of it (his mind was so full of numbers from the first three) when he suddenly realized that he could make them disappear simply by wanting them to. His previous attempts apparently had failed because he hadn't been adamant enough about wanting to forget the numbers.

There are very few examples of people like Shereshevsky, although my colleague Chris Grosskurth interviewed one for the CBC Radio series we did called "Cranial Pursuits." His name was Jacques Scarella, and he was for many years the *maître d'* of an upscale Washington restaurant. Even though the restaurant had closed nine years before, Scarella claimed to be able to select a table from the restaurant's final night and remember what the people sitting there had ordered. One was a table of six: "...two people had rack of lamb. One had the grouper, that was a lady. One had the tenderloin and didn't want any sauce. And then one had scallops ..." What was even more interesting about Scarella was that he seemed to have broken the childhood amnesia barrier—he remembered some of the people and scenery from a time when his parents had sent him to live in the mountains of Italy. He had been only six months old.

As fantastic as his memory is, Jacques Scarella is no Shereshevsky. Scarella admitted that he often forgot telephone numbers, his anniversary or his wife's birthday, but Shereshevsky's memory was so overwhelmingly powerful, so complete, that for him it became not so much memory as the nearly pathological inability to forget. His every waking moment was invaded by memories for which he had no desire or need. Shereshevsky's memory feats were amazing, but there is a select group of people who are his equal, yet suffer none of the agony of being unable to forget. These are savants, people with one outstanding mental ability—often involving memory—emerging from a background of severe disability. They are so unusual that some researchers claim we will never understand memory until we understand them.

They used to be called idiot savants, with the term savant reserved for a person of exceptional learning (although my

Funk and Wagnall's Standard College Dictionary defines savant as a "*man* [italics are mine] of exceptional learning") but idiot has been discarded as being pejorative and inaccurate. It was a term reserved for people with IQ's less than 25, but is no longer used, and most savants have IQs higher than that anyway. The original title of idiot savant did capture the paradoxical nature of these people: while mentally handicapped for the most part, they have one (usually no more) amazing skill. Sometimes it is the ability to play a complicated piece of music on the piano after hearing it only once; some are calendar calculators, and can tell you instantly what day of the week any date in history was; others are brilliant artists. There are few examples of talents other than these, which is one of the many puzzles of this condition. What they all share outside of the one sometimes unbelievable ability is a profound mental handicap, usually serious enough that they are unable to care for themselves.

Some savants have abilities that seem beyond human. Oliver Sacks reported the case of identical twin savants who could calculate prime numbers in their heads and would exchange them back and forth for amusement. (Primes are numbers divisible only by themselves and one.) Sacks himself sat in on one of these sessions, armed with a booklet listing primes, and challenged the twins by throwing an eight-digit prime into their six-digit conversation. After a lengthy (and silent) pause, they broke into smiles and came back with their own eight-digit primes. Sacks was able to confirm their correctness with his book. But when they got up to twelve-digit primes, he was lost—his book stopped at ten digits. These were two men who would have had trouble adding three and three, yet were somehow able to visualize prime numbers. The rest of us can't even grasp the concept of visualizing them let alone deal with the numbers themselves.

Dr. Bernard Rimland, who has studied savants for decades, told me a story about an individual living on the US west coast who communicates using his huge record collection. If you ask him a question, he will search through his records (I don't know if this has been updated to CDs) put one on the turntable, and drop the needle at the exact spot where the lyrics of the song contain the answer to your question.

Many of these amazing people have absolutely incredible memories. Some of the calendar calculators can not only tell you the day of the week for every date (which seems not to be a feat of memory) but can also tell you what the weather was like, providing that day was within his adult life. (I say "his" advisedly: male savants outnumber females six to one.) There was one savant in the nineteenth century who could remember the day every person in his parish had been buried, how old each was at death, and who the mourners were. There are some who argue that the calendar calculators are memory experts too, and that they simply visualize past calendars spread out before them and pick out the day of the week for the appropriate date. If that is true they might be like Sacks's twins in being able to remember and scan vast fields of numbers in their minds.

It has been suggested that those savants with fantastic memories have something in common with Shereshevsky: they cannot forget (of course a major difference was that he was not mentally handicapped). This is extremely hard to prove, but it is important to note that while being able to play back a piece of Mozart note for note could conceivably be accomplished by an extraordinary working memory, remembering who was buried in the parish requires long-term memory, a completely different system. But how do they do it?

Bernard Rimland argues that one of the keys to the savant's ability is an extremely narrow focus on certain kinds of information. He views non-savants—the rest of us—as being distracted every second by a flood of incoming information, most of it irrelevant to the task at hand, while the savant, although handicapped in one sense, is also unencumbered by the baggage of life and is free to concentrate on a single talent. My reaction to this idea has always been that I must be very distracted, because I can't possibly imagine how narrowing my focus could bring me anywhere near the levels savants attain.

What is happening in the savant's brain? Even if we forget the music and art and calendars and just concentrate on their memories, there are very few clues to what makes them fantastic. Savant memory appears to be more automatic and immediate than ours, but the way it looks to us doesn't say very much about its true nature, and most of these people are unfortunately unable to give an introspective account of their abilities. One savant investigator, Dr. Darold Treffert, suggests that savant abilities result from some damage to the left hemisphere coupled with disruption of the hippocampus and the rest of the normal memory machinery. This would leave the right hemisphere dominating the mental world of the savant, which squares with their enhanced musical and mathematical abilities. But how exactly normal memory could be *enhanced* by some disruption or brain damage is hard to visualize, unless again distractibility is reduced by allowing the memory mechanisms to focus on a very narrow range of information. Savant memory is different from ours in more than its power. As Dr. Treffert has pointed out, it is narrow but exceedingly deep, while ours, shallow by comparison, is much more wide-ranging. Theirs lacks emotional overtones, is much less flexible than ours and in some respects is unthinking in a way that resembles the kind of habit memory we rely on to drive a familiar route to work. But this is merely a list of superficial differences—so far there is no really good evidence of a different memory system at work in the savant's brain.

One intriguing incident leaves hope that one day these abilities may be understood. The famous savant calendar calculators and prime-number artists mentioned above so intrigued researchers that a psychology graduate student was recruited to see if he could learn to be a calendar calculator of their class. He memorized a formula for doing it and struggled for some time, day and night, to improve. While he could calculate the day of the week for a given date, it remained a laborious task until suddenly he discovered he could do it at savant speed. Calendar calculating has become an automatic process in his brain. Bernard Rimland speculated that this dramatic change reflected a shift in the calculations from the left hemisphere, where step by step is the *modus operandi,* to the right hemisphere, which is much better at seeing patterns as a whole. But even Dr. Rimland described this as no more than a "hunch."

We may never understand how savants' brains work nor grasp how they perform their great feats of memory and calculation. But the suggestion that they take in much less information than we do and so are able to concentrate all their powers on a relatively small amount of information may also shed light on why they don't forget in the same way we do. They don't have to. There is probably a very good reason that forgetting is a process of major importance in our brains: the important highlights of a typical day are very few in number. Most of what our brains are exposed to is the sum of the dull details of living, material well worth forgetting. The difficulty we face is that we forget the highlights too. Forgetting should be viewed more as a process of amending memories: sometimes an event is lost completely, true, but sometimes we forget an event while still carrying the memory of it. The memory has become a crudely drawn caricature of the original happening, shaped by more recent events and influences like photographs and family stories, and we are left remembering the fact that we remember, not the event itself. Some have even suggested that all we remember is the most recent version of the memory, all other recollections having been lost. You can convince yourself that this is possible by focusing on a long-held memory and then trying to expand that memory to new levels of detail. You'll find that many of your most cherished recollections are static and immovable and will resist any attempt at expansion. They are mere memories of memories.

Even if most of us wouldn't want Shereshevsky's powers of memory, it might be nice to bring back to life some of the memories we once had and have lost. This raises a question that has attracted psychologists for decades. Are memories, once lost, irretrievable? Charles Darwin's cousin, Francis Galton, was not only curious about many aspects of human nature but had the wit to ask the right questions and design experiments that would actually give answers to those questions. He is viewed with some suspicion today because he had typically Victorian attitudes towards race and class: for example he believed there was a biological basis to criminality, and devised a technique for melding together the features of a number of felons' faces in order to create composites that he hoped would reveal the typical set of criminal facial features. He was disappointed to find it didn't work.

There's no doubt that Galton was a clever experimenter even so, and he was intrigued by memory and forgetting, in particular the question of whether memories can be lost forever. Not whether any memories can be forgotten forever, because obviously most of the daily stuff in our working memories, the phone numbers and pizza combos, are gone forever a week later. Galton was asking whether or not a memory once held firmly, then forgotten, could be unexpectedly retrieved given the right cues. He experimented on himself using a set of single words (unknown to him) written on pieces of paper. He would look at a word, think casually about what it meant until a couple of ideas or vague memories occurred to him, then he would suddenly focus on those fleeting memories and try to bring them to light as vividly as possible and record them. He repeated this experiment several times, until he could begin to draw some conclusions about his stock of memories and the possibility of bringing long-forgotten ones to light.

Galton found that most of the memories stimulated by these random words came from his remote past, when he was child or a young man. Only about 15 per cent represented recent events. He also found that the same word prompted the same memory, even if its second appearance was weeks after the first. He concluded that there were strict limits to the number of recollections he was capable of and that his mind tended to run in the same grooves week after week. As for reawakening long-lost memories, he couldn't deny that it was possible, but became convinced it was an extremely rare event.

Galton's ability to introspect and scrutinize his own thought processes provided at least one example of a recollection that looked at first glance as if it had sprung out of nowhere, but turned out instead to be a fragmentary memory that was never far from his awareness. When Galton was a boy, his father had insisted he spend a few days in a chemistry laboratory to improve his general knowledge. Three times, in response to the words on cards, he remembered both the arrangement of tables in the lab and the smell of chlorine gas. As Galton himself remarked, had these two impressions been triggered in the course of everyday life, they would have appeared to be a memory recovered from nowhere, but his card experiment showed that they were hanging around in his memory banks waiting to be remembered. He concluded that "forgetfulness appears absolute in the majority of cases..." and that as fearfully dramatic as it might be, the idea that all the events of one's life might flash before one's eyes at the moment of death was just too unlikely. What he didn't add was that if that did happen it would probably just be a rehash of the same well-rehearsed memories, a truly terrible way to die.

Despite Galton's work, as recently as the late 1970s an informal survey showed that about 70 per cent of the general population (and an amazing 84 per cent of psychologists) believed that all memories are preserved and can be accessed if the right techniques are used. Psychologist Elizabeth Loftus conducted the survey and discovered that one reason people believe that all memories are permanently stored is that most of us have had the experience of suddenly recalling what seems to be a long-forgotten event. There's nothing like personal experience. Besides this conviction, a number of psychologists cited the work of Wilder Penfield. Penfield stimulated parts of the exposed brain and apparently triggered the appearance of detailed mental scenes which had all the trappings of memories, some of them decades old. As I described in Chapter 2, Elizabeth Loftus has examined in detail the testimony of these patients and finds not only that most of the descriptions are probably not real memories, but that the number of patients who had even these doubtful memories is a very small percentage of the total.

Hypnosis and psychoanalysis are also cited as techniques for bringing back that which has been hidden in the brain for years. Dr. Loftus worries that both are as likely to generate new memories—of events that never happened—as they are to recover true memories. She and her colleagues have performed a series of experiments that show just how easily that can be done.

In one, volunteers watched a series of colour slides depicting a car striking a pedestrian at an intersection. Half of those viewing this sequence saw a stop sign at the intersection, the other half a yield sign. All were then required to answer a set of questions, including one, question 17, which appeared in two versions, "Did another car pass the red Datsun while it was stopped at the stop [or yield] sign?" The experiment was designed so that some people who had actually seen a stop sign were asked the "yield sign" version of the question and vice-versa. Shortly thereafter, these people were asked to view pairs of slides and choose the member of each pair that they had actually seen previously. When it came to the critical slide, they were given a choice of a stop or yield sign, and a significant number (in one part of the experiment 80 per cent) chose the wrong slide. There are a number of possible explanations, the simplest of which is that many of these people simply didn't notice the sign at all in the original viewing, and question 17 merely supplied them with information they had never had. But this is unlikely because volunteers tested for their memory of the sign immediately after the slides—with no questions intervening—were 90 per cent accurate in their choice of the sign they had seen.

It might also be suggested that two competing versions of what happened, one derived from the slides, the other from the questions exist in the brain simultaneously. Although it is impossible to prove that there are not two memories in existence, the fact is that people will persist in choosing the wrong version (the yield sign when they actually saw the stop sign) even when presented slides of both and asked to choose between them. It's hard to explain why the wrong choice would persist in this situation if memory traces for both were still present. A variety of further tests convinced Elizabeth Loftus that questions which introduced a stop or yield sign which had not been seen not only interfered with the accurate recall of the sign, but had replaced an original memory with a false memory.

These of course were very narrowly defined laboratory experiments, but they can easily be extrapolated to situations outside the lab. Who knows what sort of questioning, prompting and encouragement a key witness to a crime has been subjected to before getting to the courtroom? If memory is so malleable, how much weight can be put on the recall of a witness who is likely under considerably more stress in the courtroom than any volunteer in a university psychology lab?

Since these experiments in the late seventies, the issue of the fragility of memories has taken on new and urgent significance for a particular kind of recall, the recovery of repressed memories of childhood sexual abuse. The 1990s have seen an epidemic of episodes of abuse being remembered twenty or thirty years after they happened, the memories usually only coming to light after the abused individual enters therapy. The theory is that the memories, traumatic as they are, have been completely repressed, with only the barest hints of their existence. These can be, according to some accounts, as simple as the unwillingness to try new things, or being unsure about what one wants. Television star Roseanne Arnold, the former Miss America Marilyn Van Derbur and literally hundreds or thousands of other people in their thirties and forties outside the public eye have remembered incidents of abuse, which might have occurred decades before, and of which they had previously been completely unaware. Lawsuits claiming damages are becoming common, and many states in the United States have introduced legislation so that victims of abuse can take legal action years after the alleged incidents happened, action that was formerly ruled out by the statute of limitations. Perhaps inevitably there has been a backlash: the False Memory Syndrome Foundation in the United States (with chapters in Canada) argues that the memories of abuse which emerge after analysis are most likely false, unconsciously created by the purported victim in response to suggestion or worse, encouragement by the therapist. Opponents, including some psychologists, argue that the Foundation is just as creative when it refers to a false memory syndrome, a term they argue has no psychological validity. This is a minefield of hatred and anger, and one that I am sure most students of human memory would like to avoid, but some, including Elizabeth Loftus, have tried to make some sense out of the flood of directly contradictory statements and claims. As a consequence of her work on the malleability of memory, Dr. Loftus fears that the very effort to recall memories that have lain dormant for decades may partially or even completely create those memories, and that such invented memories would be indistinguishable from real ones. In a review in *American Psychologist* she cites the case of a man who had originally denied accusations of child abuse, but after months of interrogation and therapy began to "remember" the events of which he had been accused, and many more, ranging from rapes and assaults to participation in a Satanic cult. Case closed? Not quite. A psychologist hired by the prosecution made up a story of an incident of abuse and told the accused that it had come from his two children. After at first denying it, the man eventually remembered the entire incident and wrote a three-page description of it. He had remembered an imaginary event. Loftus mentions that the techniques used to persuade him of the reality of the story are not dissimilar from those used to reawaken memories. But is this story really typical of what happens?

There are at this point only a few generally agreed-upon truths. One case of sexual abuse of a child is too many, and the incidence of such abuse is much higher than anyone would have admitted ten years ago. I don't know if many would go so far as Canadian writer Sylvia Fraser, who in an article in *Saturday Night* magazine described child abuse as "the gentleman's agreement whereby children are treated as sexual pacifiers, as disposable as condoms," but only those with truly bizarre standards of behaviour would deny that child abuse is evil, and that we should do whatever is necessary to eradicate it. At the same time, it is also true that the accuracy of memories which surface only after intensive one-on-one sessions with a therapist can be questioned. These two facts imply simply that some of the sudden recollections of long-forgotten incidents of abuse are true, and some are not. It's not possible to say with certainty how many belong in each category, and it doesn't appear to be possible to tell one kind of memory from the other. Beyond that simple and not-very-helpful truth is a swamp of contradictory data and interpretations. Psychologists debate the percentage of adults known to have been sexually abused as children who repressed that memory for some significant amount of time. Some psychologists even argue that repression does not exist, claiming instead that the horror of the original event prevented the details from being incorporated into memory in the first place. Generalizations simply can't be made, and each case must be considered on its own merits. But this is obviously a situation in which it is of much more than simple academic interest to clarify what happens when memories from the distant past are recalled.

The idea that our memories, the fabric from which we make our personal history, can be wrong or misleading comes as a shock at first. After all, our total set of personal memories is the only thing that is ours and ours alone. Others know pieces here and there, but only in our brains does the story unfold from beginning to end. But as it turns out, even in one's own brain the story is never completely and truthfully told.

part six
DREAMS

19
A Freudian Sleep

For nearly a hundred years the popular view of dreams in the Western World has been dominated by Sigmund Freud's idea that they are the fulfilment of forbidden childhood wishes. But the modern scientific approach spends more time trying to understand how the brain creates dreams than it does interpreting them. Most of the current research into dreaming goes on in a sleep lab, and I was lucky enough in May 1992 to spend a night in Dr. Josef DeKoninck's sleep lab at the University of Ottawa.

A sleep lab is a place where researchers expect you to sleep just the way you do at home while a wall-mounted video camera records your every movement, electrodes pull at your scalp any time you turn your head on the pillow and a long flexible plastic probe collects data on your deep body temperature through the only appropriate orifice. In my case, there was the additional chemical complication of a mid-evening meal of pizza, too much wine and espresso.

The electrodes pasted to my head were distributed around the scalp to detect brain activity, under the chin to measure muscle tone and beside the eyes to track their movements. All of these record changes in strength of the local electric field which in turn drives the back-and-forth wiggles of a pen as a roll of paper passes continuously underneath. The result of a single night's sleep can be as much as two football fields' worth of paper filled with eight sets of wavy lines.

It's simple for a sleep researcher to look at the record and pick out a moment when a dream is occurring. Periodically through the night brain-wave patterns change, each new rhythm signifying a change from one phase of sleep to another. There are four stages of sleep in which the brain waves are very different from those of waking, with stage one—the one you enter first—the lightest sleep, and stage four the deepest. The wave patterns of each different stage represent, in some still unknown way, the collective electrical activity of billions of brain cells. Every once in a while these slow waves of sleep change to a pattern much like that of waking, and at the same time the eyes start to jerk back and forth periodically. This is the now famous rapid-eye-movement sleep, and sleepers awakened from REM sleep will almost always be able to describe a dream they were just having. Although sleep experts now agree that there is also mental activity happening in the other sleep stages (so-called "non-REM" sleep) the most vivid reports—and the most "dream-like"—come out of REM sleep. My REM periods that night in Ottawa were chaotic and fragmented (pepperoni and Italian coffee beans having left their mark on my frontal lobes) but I still managed to remember three different dreams when I awakened in the morning, the most amusing, at least to the three experts who reviewed my record, being this one:

> I was returning to my van in a nearly empty parking lot when I noticed that an old friend of mine, whom I hadn't seen in two or three years, had parked her van right next to mine. The other van was bigger; mine was newer. Then she and I and some others got into my van to discuss things like price and mileage.

Allan Moffitt of Carleton University, one of the sleep researchers assigned to my case, said of this dream, "Freud might roll over." In the public mind to be "Freudian" would require sifting through that little dream story looking for sexual symbols, because we all assume that Freud thought dreams to be nothing more than elaborate disguises for sexual content. He would undoubtedly have recognized that the real meaning of the "her van was bigger but mine was newer" dream is as plain as the, uh, nose on your face.

According to Freud, the brief series of events in the parking lot were the "manifest content" of the dream, the events I experienced as the dreamer. But behind that manifest content looms the much more significant "latent content" of the dream, the network of associations and meanings that could be revealed if I were to reflect on that dream by free association, guided by a dream analyst. Who were the people in the dream, what do they mean to me and why did we have that particular conversation? These are only three of many possible questions. The latent content, once out into the open, would be much more detailed than the sound bites of the manifest content. At the base of it all, Freud would have argued, was a childhood wish that is too shameful or embarrassing for me to allow into my conscious mind; as long as I am awake it remains repressed, but it may gain access to consciousness in a dream.

However, if such repressed thoughts were to burst into consciousness in their full repulsive detail, they would at the very least ruin a good night's sleep, and so in Freud's view my mind, like everyone else's, has a twenty-four-hour censor which acts to disguise repressed thoughts as they make their way into consciousness. This means that even though the van dream could well contain these forbidden thoughts of mine, their disguises are so effective that I will be unable to recognize them until I analyze my dream. The censor is a virtuoso, using several techniques to ensure that I can't possibly glimpse my childhood wishes in my dreams. One is condensation, the act of taking the complex sets of associated thoughts existing in the unconscious and editing them into a few minutes of dream time. My whole van dream seemed to be less than a minute, but thanks to condensation might contain hours of self-revelation. Sometimes condensation can mean that one dream figure or place actually represents two real ones. The censor also can trick the dreamer by making innocuous objects or people seem important and vice-versa. Was my emotional attachment to my old friend or to my van? Or to her van? Or to Van Morrison?

The censor also makes abstract ideas concrete; fear, jealousy or anger all must show up on the dream stage in concrete terms, as some sort of visual image. And last but certainly not least, the censor performs its best-known work of using symbols for sex. An umbrella is a penis, an oven is a vagina and a van is likely a little of both. At this point a slightly more detailed version of my dream will illustrate just how Freud's concept of the dream has become so popular (if at the same time somewhat unfaithful to the original) that any of us can turn an apparently innocent story into a sordid romp through the unconscious. Here is the van dream with some further details. I have also taken the liberty of putting exclamation marks wherever Freud's eyebrows might have raised:

> I had parked my van(!) in a large(!) parking lot (!) situated below grade(!!), such that you had to drive down a shallow hill into it(!!). I was returning to it when I noticed another one had parked immediately(!) beside it, and in that van were this old friend of mine(!!) and her boyfriend(!!!). I didn't know him but he bore a close resemblance to someone who had been in the sleep lab that night. Anyway, he was very keen to drive my van(!!)—it was newer than his (!!!)—and so the two of them and I got in(!!!!) and there was a lot of discussion about how much it cost(!!) and so on.

This was one of three dreams I managed to remember the morning after in the sleep lab, and my feeling was that this one had come from an earlier time in the night. Freud argued that the same subject is dealt with by all the dreams in a single night, so while my other dreams focused on apparently disparate subjects like reading my own sleep record the next morning (not an unusual dream the first night in a sleep lab) and walking towards an old brick high school where the students were driving by in cars from the 1950s festooned with flowers, dream analysis would have sought the common ground among them. Note that the appearance of an individual whom I'd seen in the sleep lab hours before is not a problem for Freudian analysis. Freud explained that the reappearance in a dream of events or people from the previous day resulted from the pairing of that so-called "day residue" with the unconscious repressed wish.

The research that emerges from sleep labs such as the one I was in has loosened Freud's grip on dreams. In particular, the approach today is built on data-gathering rather than intuition, the conclusions determined by statistical significance, not the creative impulsive of a single person. I would not argue that this is always a better thing—some books detailing the most recent experiments on dreaming are surprisingly tedious given the general fascination with the subject, and Freud's approach to dreams can provide enlightenment. But the experimental approach has nonetheless succeeded in picturing dreams as something other than Freudian. For instance, cats and rats seem to have dreams too, and the chemical and electrical events occurring during their dreams are paralleled very closely in ours. If dreaming (or at least REM sleep) is pretty much the same physiologically among such distantly related animals it must be a very ancient brain activity, maybe more than a hundred million years old. This notion has persuaded a number of researchers of two things: that the Freudian approach might be reading too much into something that is not uniquely human, and that it is worthwhile looking for other plausible reasons why we might dream.

It was not until the early 1950s that REM sleep was even discovered, and it took another decade or so to establish just how odd this state of mind is. The vivid dreams of REM sleep are not the strangest aspect of this period of the night. A simultaneous drop in muscle tone in the chin indicates that a general body paralysis sets in during REM sleep. You can wiggle a toe or the thumb; sometimes the fingers might jerk spasmodically, but essentially you are incapable of moving. Some people may have experienced a dim awareness of this in the first moments of wakening out of a dream (and it's even been suggested as the physiological basis for the frightening dream in which you're trying to run and can't) but most of the time the paralysis has vanished by the time you wake up sufficiently to be aware of it. The one clear exception is the sudden terrifying attack of sleep paralysis that a person with narcolepsy may experience.

Narcolepsy is a neurological disorder causing excessive sleepiness: narcoleptics run the risk of dozing off at work or while driving, and most of them take stimulants to ward off sleep. They may also suffer from a set of odd symptoms that make sense only when you view them as dissociated fragments of REM sleep. The times just before falling asleep or waking up are the worst for narcoleptics, because it is then that they enter the world of dreams while fully awake. They may experience vivid hallucinations just on the edge of sleep called "hypnogogic illusions"—one narcoleptic told me that if she lay in bed and stared at the ceiling light fixture, snakes would start writhing out of it. At the same time, these people may experience sleep paralysis, the immobility that we always have during dreams but are rarely, if ever, aware of. Some narcoleptics can lie in bed paralyzed but fully awake for up to ten minutes. But stranger still, a narcoleptic may collapse into total sleep paralysis while awake and even standing up. Expressing some strong emotion or laughing loudly may precipitate these attacks. Narcolepsy is treated with a variety of drugs that interfere with or mimic neurotransmitters, but the exact cause is unknown.

Why should one be paralyzed during dreams? It might be

self-protection: cats whose sleep paralytic centre is inactivated enter REM sleep as usual, but then instead of just lying there twitching they suddenly get up, stalk some imaginary prey or bare their lips as if they are acting out a set of dream actions. If the same thing were likely to happen in humans, we are probably safer being immobilized. The third element of REM sleep is the eye movements themselves, the subject of great controversy. Are they simply movements driven by some sudden heightening of brain activity, or are they literally following the events of the dream? There have been many attempts to establish which of these is true, and while some intriguing reports have surfaced (like eyes bouncing up and down repeatedly as the dreamer climbs a set of stairs), there doesn't appear as yet to be conclusive evidence that specific eye movements reflect changes in gaze in the dream story. It might even be the other way around, that eye movements precede and dictate, if not the details of the story, at least the abrupt changes in narrative flow.

Another feature of dreams we all share is their regularity. Ninety minutes pass between REM periods; there are four or five of them a night, and while the first REM period might be only ten to twenty minutes, each successive one is longer, with the result that you can easily spend two hours in REM sleep every night. This pattern holds for just about anyone who doesn't have a serious sleep disorder. Rhythmicity like this implies the existence of biological clocks which switch between sleeping and waking and among the various stages of non-REM and REM sleep. The scheduling of dreams is biology not psychology, and some researchers have been willing to go much further than that and argue that the Freudian interpretation of dreams, based as it is on single dream reports interpreted by Freud himself, is simply unscientific and should be abandoned. Some have even gone so far as to assert that dreams are meaningless. There are other researchers who, while acknowledging the biological nature of dreams, are still willing to accept that they contain messages that can be analyzed and are important to the dreamer. But there's no doubt that the discovery of REM sleep and its accompanying suite of features set dream science off in a whole new direction.

Even given the similarity of the dream experience for most of us, there are some weird variations in the dreams themselves. One is "lucid" dreaming, the kind of dream in which you become aware that you are dreaming but don't wake up. You continue the dream story, and can sometimes even direct it. Some people have trained themselves to be lucid dreamers; one method can be used in the early morning if you wake up out of a dream. The strategy is to wake up fully while holding the dream in memory, engage in some vigorous and demanding activities like reading and/or pacing up and down then lie down again. At this point you rehearse the dream and imagine yourself returning to it. If you do return to the dream, you might at the same time be aware that it is a dream. Another more difficult technique is to practise asking yourself repeatedly during the day, "Am I awake, or is this a dream?" Once that habit is established, it will occasionally occur to you during a dream and you might then be able to pick up enough clues to decide that you are indeed dreaming. At that point you have reached lucidity. It has obvious benefits for people who are plagued by nightmares, and offers a unique stage for people who like to fantasize. Most people have had dreams of flying—imagine that dream with the important difference that you are in complete control of the flight. (Note that a lucid dream of flying would be almost exactly like an out-of-body experience, except that few lucid dreamers see themselves floating above their own bodies).

Lucid dreaming is a rare intrusion of waking consciousness into a dream—in a sense the opposite of narcolepsy, the intrusion of a dream into consciousness—and its confirmation has led to the recognition of yet more versions of this combination. One common one is "false awakening" in which you dream that you're waking up, getting out of bed and getting dressed only to wake up out of this dream and find you have to do it all over again. Celia Green of the Institute of Psychophysical

Research in Oxford, England has suggested that falling into a lucid dream unexpectedly during the day—if it ever happens—could account for a whole range of strange "sightings," like ghosts. The ghost in this case would not be an apparition superimposed on the real world, but merely one figure in an entirely dreamed-up scene. One accomplished English lucid dreamer, Alan Worsley, has succeeded in forcing himself into a lucid dreaming state directly from waking. It sounds simple: he just lies still. Unfortunately such dreams, although attainable, are long in gestation and short in duration: it can take two hours or more for them to first appear, and when they do they are often so vivid that it's too difficult to keep still. Moving ends the dream. Worsley put it best himself: "If one dreams, as I have, in rich tactile and auditory imagery, of being examined in the dark by robots or operated on by small beings whose goodwill and competence may be in doubt…it can be very difficult to keep still." Lucid dreaming strongly suggests that the boundaries between sleeping and waking are not rigid.

Myths about dreaming abound. One is that if you dream that you are dying, you really are dying. Not true. Another is that dreams take only an instant, even if the story they tell would take many minutes in real life. This also seems not to be true, although the source of this myth is this fascinating dream about the French Revolution by a nineteenth-century dream researcher, Alfred Maury:

> I dreamed of the Terror; I was present at scenes of massacre, I appeared in front of the revolutionary tribunal, I saw Robespierre, Marat, Fouquier-Tinville, all the ugliest faces of that terrible era; I spoke with them; at last, after many events which I remember only vaguely, I was judged, condemned to death, taken by car to an immense concourse on the site of the Revolution; I mount the scaffold; the executioner binds me to the fatal plank; he makes it seesaw; the blade falls; I feel my head separate from my body…"

Maury woke up at this very moment to find that his bedrail had fallen on the back of his neck, in exactly the place that would have been targeted by a guillotine blade. Maury reasoned from this that the entire set of events in the dream must have been created in the moment or two between the time he had been struck and waking, and he therefore concluded that even such elaborate dreams might take only a second or two. But there have been many studies since which have attempted to establish the time dreams take, and most suggest that dreams take no less time than the real events would. One way of investigating this is to waken people at different times after a REM period begins and ask them to recount the dream story to that point. If dreams unfold in real time, the longer the delay before waking, the more details should appear in the story, and there should be some correlation: an extra five minutes of dream should allow for roughly five minutes' worth of events. Both are true, and so most dream researchers assume that dreams take place at the same pace as real life. Maury's dream remains unexplained, although it is well established that a dream that is already underway can easily incorporate sudden and unexpected events, like the sound of a bell or the feel of a spray of water. Such extraneous events can be made to fit the story of the dream in a way that is at least as logical as the rest of the dream. In one study almost half the dream reports from people who had been sprayed by a fine jet of water while in REM sleep contained references to falling water. However, it seems pretty farfetched to suggest that Maury had a series of fragments already in mind that were instantaneously assembled into a story about the French Revolution at the prompting of the fall of the headboard. There is no shortage of other explanations for Maury's dream: maybe the bedrail creaked in warning, giving his mind time to run through the dream events before it fell; maybe the smooth flow of this dream narrative is more a product of his storytelling ability than a literal account of the dream; or maybe this was a dream of precognition, a prediction that the bedrail would fall. Some dream investigators have gone so far as to suggest that the report is not reliable to begin with, which of course is the easiest way to explain it. However, Maury's mother was at the bedside (her son was an invalid) and attested to the instantaneous nature of the dream. But how would she have known how long his dream ran?

All of us have experienced the intrusion of a real event into a dream story, and this shared aspect of dreaming is one of the reasons it is so fascinating. Dreams seem so personal and idiosyncratic (who else could assemble the same group of people and events?) but there are intriguing points of commonality at many levels beyond the purely physiological. One is that the first dreams of the night tend to deal with the most recent events, while later dreams dig further into the past for their script. Another is that the settings for a dream and the characters and events played out in those settings might come from two completely different eras in your life. Many people discover the same setting from childhood, like a house and backyard, appearing over and over in dreams whether that setting is appropriate from the point of view of the story or not. But the two most important common features of dreams are that they are bizarre and easily forgotten. A dream theory must be able to explain both of these features to be worth anything.

If dreams are important enough to occupy the brain four or five times a night for extended periods of time (if not continuously through all phases of sleep), why is it quite common not to remember a single dream for several nights in a row? Although it's not possible to know for sure, there has been at least one estimate that 95 per cent of all dream content is forgotten. Freud had a ready explanation for dream amnesia: the damaging nature of the infantile wishes contained within the dreams meant that they had to be returned to the unconscious and repressed again as soon as possible—the genie had to be put back in the bottle. Today's non-Freudian theories about why we dream must also come to grips with this difficulty in remembering them.

The other aspect of dreams that demands explanation is their bizarreness. I happen to think my van dream was pretty straightforward and self-contained, but it could be that it was really only a fragment of a larger, much more disjointed dream. It wouldn't have surprised me if the parking lot had suddenly become a market, or if the people who got into my van kept changing every time I turned my head or if I had suddenly found myself in a restaurant. I'm certainly capable of odd dreams: I have had a recurring dream of trying to play either squash in a court full of furniture, or basketball in a hardware store. The unpredictable and unnatural nature of dreams is what sets them apart from thinking while awake.

Freud attributed bizarreness to the contortions of the dream censor: how could a dream not be bizarre after the censor has condensed the story, made abstract ideas concrete, shuffled people and objects so as to disguise the true source of emotional conflict and then loaded the whole dream with sexual symbols? Again, as we will see in the next chapter, today's dream theorists agree that the idea of bizarreness in dreams must be addressed by any decent theory, but few will attribute it to a Freudian dream censor.

20
Dreams of Echidnas

I dreamt recently that my next-door neighbours were cooking in their kitchen using some sort of powerful overhead light and the food had begun to burn. We all rushed inside and I found myself in the basement admiring the job they had done of cutting deep channels through the concrete floor and filling them with water so their children could play with toy boats. The next thing I knew, I was outside talking to a friend of mine using what appeared to be a child's cellular telephone. Just as I began to worry that I had been a little too sarcastic in imitating my friend's English accent, a beautiful kite appeared overhead and I was able to comment on it and change the subject.

This is a perfectly run-of-the-mill dream in that it jumps from subject to subject with no apparent reason and juxtaposes the unlikeliest of scenes without missing a beat. No one would be surprised that I had a dream like that, nor that I, as the dreamer, accepted it all unquestioningly. On the other hand, that dream is atypical because I remembered it. We've all had the experience of a fleeting dream image that disappears a few moments after waking, but what we don't realize is that most dreams are forgotten completely. Any dream theory today has to make sense of these dream characteristics and at the same time incorporate—or at least acknowledge—forty years of research into the electrical and chemical nature of brain cells. That doesn't mean that today's researchers are strait-jacketed by the facts—there's still lots of opportunity for creativity in matching the characteristics of dreams, especially their bizarreness and forgettability, to EEG's and neurotransmitters. A sampler of dream theories will show you what I mean.

One idea that made a big press splash when it was first suggested about a decade ago came from scientists Graeme Mitchison and Francis Crick, the co-discoverer with James Watson of the structure of DNA. They contended that the forgetting of dreams wasn't just a side-effect, but indeed was the whole purpose of dreaming: "We dream in order to forget" was their phrase. They argued that dreams have no significance whatsoever because they are just glimpses of material that the brain is dumping in a nocturnal purge of excess information. Crick and Mitchison based their argument on computers rather than living things. In particular, they pointed out that neural nets—simple computer-based networks designed to mimic the workings of assemblies of real brain cells—encounter big problems if they are inundated with information. They can even be forced into abnormal processing that mimics human mental behaviour, like coming up with inappropriate associations among bits of information (fantasy) or getting locked in endless repetition (obsession). What a neural net in this plight needs to do, according to Crick and Mitchison, is to give itself a tune-up by shutting down all inputs and outputs, stimulating itself repeatedly from within to excite these rogue responses, then damping them out as they appear. Funnily enough, this tuning-up process—especially the processes of isolation and self-stimulation—is reminiscent of REM sleep and dreaming. They called the process of ridding the brain of unwanted information "reverse learning."

If Crick and Mitchison are right, remembering a dream

would actually be undesirable, the equivalent of seeing incriminating documents just before they disappear into the shredder. But there are problems here: according to their theory, a dream once had should be gone forever, but all of us are familiar with recurring dreams. For people who are suffering through emotional crises, they are the norm, not the exception. Crick and Mitchison were forced into proposing that waking up in the middle of a dream causes the act of dumping to shift into reverse, ensuring not only that the dream is remembered, but that it will return.

As a sidelight to their theory they raise the enigma of the spiny anteater, the echidna, which lives only in Australia. This is a primitive forest-dwelling mammal, a relative of its fellow Australian, the duck-billed platypus. It appears to be the only land-dwelling mammal that doesn't have REM sleep. But it does have a huge set of cerebral hemispheres, especially the frontal lobes, even bigger in relation to the size of the animal than ours. Crick and Mitchison suggest that it needs that huge brain because its lack of REM sleep makes reverse learning impossible, leaving the echidna incapable of ridding itself of surplus information. Its only recourse is to store that information somewhere in its huge cranium, and here the computer theory agrees: if you can't tune a neural net, the next best thing to do is expand it.

Crick and Mitchison aren't the only dream theorists to have taken note of the echidna. (It says something about dream theories of the eighties and nineties that the spiny anteater makes as many appearances as Freudian wish fulfilment.) Dr. Jonathan Winson of Rockefeller University finds support in the echidna for his dream theory too, even though it is almost exactly opposite to that of Crick and Mitchison. Rather than "dreaming to forget," Winson sees dreaming as a way of remembering information that is most crucial to the survival of the animal, whether rat or human.

Winson's theory of dreaming mixes chemistry and electricity with behaviour. Any animal that is busy doing something that is essential for survival—like stalking or exploring—shows an unusual pattern of brain waves. Blow the odour of cabbage into a drowsy rabbit's nostrils and its brain waves change abruptly, especially in the hippocampus, the part of the brain renowned for its importance to memory. There the near-sleep profile of large jagged waves gives way immediately to a regular, six-per-second pattern called "theta" waves. The behaviours that trigger theta vary: for rabbits it might be the alertness triggered by suspicious sounds (not to mention cabbage); for cats, stalking; and for rats, the exploration of their surroundings. (Exploration is so important for a rat that a hungry one will examine its surroundings before eating food that has been put in front of it.) Given that these waves show up in the hippocampus, it seems reasonable that theta waves should play some important role in remembering the salient points of these behaviours.

Winson was able to provide support for this conjecture by recording the electrical activity from single brain cells in the rat's hippocampus. As incredible as it might seem, during the exploration of a maze the rat's brain assigns the memory of certain locales to individual brain cells. There may be more than one cell available to record the same important place, but the remarkable thing is that a single cell can do it at all. Winson and his colleagues were able to find some of these "place neurons" in the rat's brain and show that they fired rapidly only when the animal was in a certain place in a maze, for example, just inside the entrance to passage number 1. As soon as the rat wandered away from that location, that place neuron stopped firing, and others began. Presumably every significant location in the maze was tagged in the rat's brain by a place neuron.

Then came the connection to dreaming. Theta rhythms also show up in the hippocampus during REM sleep. Jonathan Winson was able to show that these same brain cells that are sensitive to a particular place in a maze are also highly active during REM sleep. He concluded that during REM sleep they are being reactivated in order to commit to memory their record of the rat's explorations. Those memories require changes in the chemistry and architecture of the receptors at the synapse, raising the problem of how to bridge the gap between the electricity of theta waves and the synapse. One is a process, the other a structure—how do they work together to record the day's events? It turns out that, for reasons unknown, the physical changes at the synapse occur best in the presence of theta waves. So theta waves turn on at important times while an animal is awake, then are replayed during REM sleep while those events are being consolidated into memory. Jonathan Winson argues that REM sleep, at least in these animals, is a time when memories of those behaviours most important to survival are brought back to life and committed to long-term memory, or as he put it to me, "The cat has to relive the event to remember it."

And the echidna? Its hippocampus produces theta waves too, especially when it is rooting around in the soil searching for its insect prey. But it lacks the REM sleep that (at least according to this theory) would allow it to replay and consolidate memories of those activities. So Winson suggests the echidna has huge frontal lobe convolutions in order to store permanently all the crucial information it gathers during the day as it comes in. (This is the exact reverse of Crick and Mitchison's notion that the big brain is necessary to store *useless* information that can't be gotten rid of.) Animals with bigger brains than the echidna couldn't afford to have frontal lobes of echidnalike proportions or their brains would have had to be grotesquely big—Winson likes to say that humans would have needed wheelbarrows for their brains. That is why (if you buy this theory) the appearance of REM sleep in evolutionary history is so important: it allowed mammals to allocate space in their frontal lobes for complex information processing instead of requiring them to serve as giant neural parking lots.

But this theory has its problems too. For one thing, theta waves have not yet been discovered in the brain of any primate, including humans. Winson doesn't seem terribly concerned by their absence. He points out that at least the changes in synapses necessary to create long-term memories are occurring, and it must be that they are facilitated by some different, and as yet undiscovered brain-wave rhythm. In his view it would be likely to be faster than the six-a-second theta, because while theta was perfect for animals which depended primarily on their sense of smell, a sense that needs only moderate computing power, humans and their primate relatives depend on vision, a sense that became dominant later in evolution and one which relies much more heavily on speed.

Can this model of dreaming-to-remember be applied to the bizarre and fascinating content of human dreams? It can if you take the word "survival" in its broadest sense. The human equivalents of being on the alert for danger, seeking prey and exploring could be insecurity, fear, jealousy, anger, even Freud's frustrated childhood wishes. Winson goes so far as to suggest that the brain systems involved in using REM sleep to reprocess information gathered during the day are collectively the unconscious as Freud defined it. Winson believes that dreams are not well remembered because they don't have to be—the rerecording of these events in memory is going on whether you remember the dream or not. On the other hand the process isn't disturbed if you do remember a dream.

While there is not yet evidence of theta rhythms in the human hippocampus, there is growing evidence of a connection between REM sleep and learning in humans. Carlyle Smith of Trent University in Peterborough has shown that if students spend time learning tricky logic tests, there are changes in their REM sleep in the following nights. The surprise is that word "nights," because in these students (as in lab rats) periods of learning difficult material are followed by REM sleep changes as much as seventy-two hours afterwards. In one experiment students learned the Wiff'n'Proof logic game between 4 and 6 P.M. and were then divided into groups. Some stayed up all that night, some slept that night but stayed up the next night, and some stayed up the third night. The results were strange: sleep deprivation on the first and third nights impaired subsequent performance on the game, but staying up the second night had no effect. Smith went on in subsequent experiments to show that being wakened repeatedly out of non-REM sleep had no effect on students' ability to play Wiff'n'Proof the next day, but being wakened out of REM sleep did. Smith argues that there are periodic REM "windows"—some of them days after the original events—during which learning and memory are somehow dependent on REM sleep. The fact that these windows skip the second night seems to me to create the perfect opportunity for learning two things, say Wiff'n'Proof and chess, on successive days then having the brain see-saw back and forth in the next few nights, first getting Wiff'n'Proof down, then rehearsing some unusual chess openings, then back to Wiff'n'Proof and so on. At the very least, these experiments suggest that learning and REM sleep are somehow connected.

Jonathan Winson's theory of dreams and learning captures both the pluses and minuses of current dream research: it has moved far beyond Freud in its ability and willingness to take the actual workings of the brain into account, but still falls short of being able to account for the richness of dream imagery and experience. The problem with analyzing dreams scientifically is that they are completely personal. When I relate a dream to you, you have to trust me: you have no idea whether I am telling the truth, the whole truth or something else altogether. I may not even know if I'm relating the dream correctly. You might infer from my brain waves that I was dreaming, but that's as far as you can go. Freud got around the problem by being unscientific and anecdotal, but he told a good story. Unfortunately the search for the brain basis of dreams isn't nearly as compelling, because it doesn't address individual dreams. What is more titillating: knowing the forbidden wishes behind a dream or identifying receptors which are active in the hippocampus during REM sleep?

While Crick and Mitchison's and Winson's theories offer

handy, albeit completely opposed, explanations for why we forget dreams, one of the dominant theories over the last fifteen years has concentrated instead on using the biology of the brain to explain the bizarreness of dreams. Allan Hobson of Harvard is the man behind the theory. In 1977 he and Robert McCarley put forward the suggestion that dream stories were simply the brain's best effort to make sense of a randomly generated series of visual images stimulated by activity in the brain stem. Their original argument was that dreams, far from being fraught with meaning, meant exactly nothing.

Hobson thinks that the strangeness of dreams (the abrupt changes of people and places, incongruous scenes and proliferation of *non sequiturs* and ad hoc explanations) results from two distinct chemical changes that occur during REM sleep. One is the shutdown of key neural centres in the brain stem (the part just above the spinal cord) which causes a dramatic depletion of two neurotransmitters normally supplied to the frontal lobes by those centres. Because these transmitters play an inhibitory rather than excitatory role, their loss releases inhibition, and makes unpredictable mental activity more likely; the bizarre is allowed to run free. Coupled with that is the eruption of sudden and random electrical discharges called PGO (pontogeniculooccipital) spikes from another brain stem centre. In Hobson's view these help drive the rapid eye movements associated with dreaming and act on the thinking processes of the dreaming brain in much the same way that a surprise event captures our attention when we're awake. Any animal that hears a sound off to one side or sees the sudden appearance of a shadow shifts attention immediately in that direction, and in the brain that shift is marked by a burst of PGO waves. The only difference during dreaming is that there is no outside event—the PGO waves burst of their own accord. In a sense then, the depletion of brain chemicals makes the brain unstable, and the PGO waves are enough to send it over the edge. The result is the disjointed world of dreams.

Allan Hobson is a dominant figure in dream research, if only for his prolific publishing: in one recent issue of a journal called *Consciousness and Cognition* all eight reports were authored by him and his team at Harvard University. Whether because of this or despite it, few of his views are given unanimous assent by his colleagues, including the explanation for the bizarreness of dreams. Even though we've all had dreams that seem completely weird, there are researchers who dispute the very existence of bizarreness in dreams. Jonathan Winson (of the "dream to remember" theory) contends that while there might be some odd associations in the images created during dreams, bizarreness is simply a "decoding problem": dreams only seem peculiar when we can't figure out what they're trying to tell us. It might be true that understanding a current source of worry might suddenly make clear the associations among a strange set of dream images, but Hobson sees additional incomprehensibility in the way the dream story is put together, in the sudden changes of scene and juxtapositions of weird ideas. When it comes right down to it, Hobson is much more prepared to deny meaning to a dream than is Winson.

But Winson isn't alone in wondering about bizarreness. John Antrobus is a dream researcher at the City College of New York who has made some surprising discoveries, one resulting from the shift forward by three hours of volunteers' time of going to bed and getting up, so that they end up dreaming their last dreams of the night at a time when they would normally already be awake. In these morning hours the brain is already reaching near-waking levels of activation—it is emerging from sleep—and dreams reported from this time are stranger and more disjointed than dreams from earlier in the night. No one knows why, although one suggestion is that an active brain will be more aware of the bizarreness of its own dreams. Some parts of the brain might literally be paying closer attention to the shifts in attention and narrative produced by other parts. If the idea that some parts of the brain might be listening to or watching what other parts are doing seems strange, remember lucid dreaming, in which a dream scene is being played out while you, or whatever part of the brain is representing the "you," look on, and may even intervene. John Antrobus wonders if bizarreness in dreams says more about the state of the brain and how activated various regions are than on the chemical switches Hobson favours.

An even more striking observation made by Antrobus is that dreaming may not be the only time our minds wander off the beaten track. He compared the content of dreams with the stray thoughts of people who were left alone lying down in a darkened room and were later asked to "tell everything that was going through your mind before I called you." Their daydreaming was, at least on the measures Antrobus used, *more* bizarre than the dream stories. For example, discontinuities, incidents usually prefaced in the dream report by the phrase, "and all of a sudden..." were twice as common in waking as in REM sleep. As Antrobus points out, conventional comparisons of dreaming with waking may be too simplistic. While we're awake our brains may be required to carry on a conversation (or worse, an argument) during which close attention has to be paid to a logical development of ideas; at other times we might allow our brains to idle. Antrobus's results with people free-associating in a dimly lit room suggest that the idle brain is perfectly capable of being bizarre even in the absence of any EEG evidence that dreaming has occurred.

He suggests that it makes sense to think of both a visual and a conceptual centre at work here. When one is awake, the visual centres pass on the images they have assembled and the conceptual centres interpret and make sense of them. The same thing happens in REM sleep, except that the vision areas are somehow generating their own images in the absence of actual seeing. The problem, according to Antrobus, is that these vision areas in the brain, while they are great at assembling shapes, colours and shading into visual images, haven't a clue as to how to put them in a sensible order, because in normal waking life that is done for them by the eyes tracking events as they happen. There is a temporal order to things, and the visual system simply preserves it. In sleep there is nothing dictating that order, and so images may be put together in strange ways. (As an aside, it shouldn't be forgotten that the dreams of REM sleep may be only part of the story, albeit the most vivid part. There is good evidence that the sleeping brain is active at other times: People awakened from non-REM sleep often report that thoughts have been running through their minds, although the thoughts have none of the strange and vivid character of those from REM sleep. There is even the suggestion that dreaming occurs throughout the day as well, but is obliterated by the flood of sensory information that characterizes waking.)

So much sensory data feeds into the brain during waking that no attention is paid to most of it. And even information that isn't screened out by the sense organs themselves is held in short-term memory only for a matter of moments before being allowed to disappear. Sensory input even directs the development of the infant's brain—a child born without sight or hearing ends up with a much different brain structurally than one whose sensory inputs are complete. So tied to the interpretation of sensation is the brain that when left without it, it creates its own. John Antrobus's daydreaming volunteers, people floating in sensory deprivation tanks, and all of us when we dream experience just what our brains are capable of if left to write their own scripts. Even though almost all sensory information is absent (the exception being the sudden sound that may intrude into the dream story), those parts of the brain whose job it is to interpret, find patterns and make sense of images are alert, functioning and capable of creating a story out of anything that comes their way. Whether the material for that story is repressed childhood wishes or random firings by brain cells is irrelevant to the storyteller. It's a doubly amazing world because the dream state suspends our disbelief so that nothing that happens in a dream surprises us as dreamers.

Freud called dreams the "royal road to the unconscious," but they are also a sampler (albeit with the appearance of having been assembled by a demented rock video producer) of what the brain is doing during the day: recalling memories, navigating through maps of space, recalling familiar faces and even weaving exotic stories. The significance of dreams is not just that they are strange—it is that the strangeness we find so peculiar to dreams is never far from the surface even in the waking brain.

An example is the burning-house experiment, which demonstrates that much can go on in our brains that is outside of our awareness. By that I don't mean the relatively trivial business of processing and packaging visual or auditory signals, but rather the elaborate storytelling and rationalization that is fundamental to our mental make-up and personality, even though it is based on thinking that never enters our "minds." Lest anyone think this unawareness is confined to people who have suffered brain damage, remember the panty-hose experiment, where volunteers invented reasons for their choice which clearly had no factual basis. How strong can the contrast be between the so-called "rational" mind and the dreaming brain when both are busy assembling streams of thought out of reach of our consciousness? You have at least two brains (and probably many more): one produces the thoughts you think you control, the one you are likely referring to when you "make up your mind"; the other is the one that—unbeknownst to you—is making up your mind for you. All the more reason to sit back and enjoy your dreams: they are giving you a behind-the-scenes glance at the inner workings of the brain.

Additional Reading

Chapters 1, 2 & 3

Ornstein, Robert. *The Evolution of Consciousness*. New York: Prentice Hall, 1991. Although this book is specifically about consciousness, Ornstein writes about the brain with clarity and humour - two rare qualities.

Luria, Aleksandr. *The Man with a Shattered World*. Cambridge: Harvard University Press, 1987. This book is a devastating account of what it is like to be brain-damaged. If one had to choose a single book on the brain, this might be it.

Gregory, Richard, ed. *The Oxford Companion to the Mind*. Oxford: Oxford University Press, 1987. If you are serious about wanting to know more about the brain - for years to come - this is the book to have.

Chapter 4

Bisiach, Edoardo, and Claudio Luzzatti. “Unilateral Neglect of Representational Space.” *Cortex* 14 (1978): 129–133. The cathedral square experiment.

Guariglia, C., A. Padovani, P. Pantano and L. Pizzamiglio. “Unilateral Neglect Restricted to Visual Imagery.” *Nature* 364 (15 July 1993): 235–237. The man who exhibited neglect when looking at a scene, but none when calling up mental images.

Grüsser, O-J., and T. Landis. *Visual Agnosias.* Volume 12 of *Vision and Visual Function*, edited by John Cronly-Dillon. Boca Raton: CRC Press, 1991. This book, academic as it is in tone, is also a collection of memorable stories, including the neglect patient who attributed his troubles in traffic to "rude Berliners," the anonymous scientist who had an OBE, the woman who saw poodle faces in the subway and the lone climber who was misled by a "presence."

Chapter 5

Halligan, Peter, and John Marshall. "Left Neglect for Near But Not Far Space in Man." *Nature* 350 (11 April 1991): 498-500. The dart player whose neglect faded with distance.

Schilder, Paul. *The Image and Appearance of the Human Body.* London: Keegan Paul, 1933. Schilder's book is out of date, but he had no shortage of interesting ideas.

Corbetta, Maurizio, Francis Miezin, Gordon Shulman and Steven Petersen. "A PET Study of Visuospatial Attention." *The Journal of Neuroscience*, 13 no. 3 (1993): 1202–1226. The study picturing the attention centres in both hemispheres.

Battersby, William, Morris Bender, Max Pollack and Robert Kahn. "Unilateral 'Spatial Agnosia' ('Inattention') in Patients with Cerebral Lesions." *Brain* 79 (1956): 68-93. The man talking about his own neglect is quoted in this paper.

Baxter, Doreen, and Elizabeth K. Warrington. "Neglect Dysgraphia." *Journal of Neurology, Neurosurgery and Psychiatry* 46 (1983): 1073-78. The gentleman farmer who couldn't read the left side of his "inner screen."

Chapter 6

Marshall, John, and Peter Halligan. "Blindsight and Insight in Visuo-spatial Neglect." *Nature* 336 (22/29 December 1988): 766–767. The original burning-house experiment.

Bisiach, Edoardo, and Maria Luisa Rusconi. "Breakdown of Perceptual Awareness in Unilateral Neglect." *Cortex* 26 (1990): 643–649. The extension of the burning-house experiment to include chipped wineglasses and vases with flowers.

Wilson, Timothy De Camp, and Richard Nisbett. "The Accuracy

of Verbal Reports About the Effects of Stimuli on Evaluations and Behavior." *Social Psychology* 41 no. 2 (1978): 118-131. People choosing panty hose for all the wrong reasons.

Gazzaniga, Michael. "Organization of the Human Brain." *Science* 245, (1 September 1989): 947-952. The left-brain interpreter is dealt with here.

Chapter 7

Grobstein, Paul. "Directed Movement in the Frog: Motor Choice, Spatial Representation, Free Will?" In *Neurobiology of Motor Programme Selection*, edited by Jenny Kien, Catherine McCrohan and William Winlow, 250–279. Oxford: Pergamon Press, 1992. Frogs' perception of the compass direction of prey is run by different brain circuitry than is the estimation of distance and height.

Flanders, Martha, Stephen I. Helms Tillery and John F. Soechting. "Early Stages in a Sensorimotor Transformation." *Behavioral and Brain Sciences* 15 (1992): 309–362. Psychology students show similar divisions of labour in their brains as do frogs.

Goodale, M.A., J.P. Meenan, H.H. Bülthoff, D.A. Nicolle, K. Murphy and C. Racicot. "Separate Neural Pathways for the Visual Analysis of Object Shape in Perception and Prehension." *Current Biology* August 1994. In press. Describing an object and being able to grasp it are two different things.

Chapter 8

Libet, Benjamin. "Subjective Antedating of a Sensory Experience and Mind-Brain Theories." *Journal of Theoretical Biology* 114 No. 4 (1985): 563–570. Libet answers critics of his work.

Behrman, Marlene and Morris Moscovitch. "Object-Centered neglect in patients with Unilateral Neglect: Effects of Left-right Coordinates of Objects." *Journal of Cognitive Neuroscience* 6 No. 1, 1994, 1-16.

Lackner, James. "Some Proprioceptive Influences on the Perceptual Representation of Body Shape and Orientation." *Brain* 111 (1988): 281-297. The experiments showing that people can be convinced that their noses are thirty centimetres long.

Chapter 9

Lacroix, Eric, Ronald Melzack, Deanne Smith and N. Mitchell. "Multiple Phantom Limbs in a Child." *Cortex* 28 (1992): 503-507. The girl with three phantoms.

Melzack, Ronald. "Phantom Limbs." *Scientific American* 266 (April 1992): 120-126.

Chapter 10

Hohol, M.J. and Sandra E. Black. "Asomatagnosia: Denial of Ownership of a Paralyzed Limb." *Neurology* 42 Supp. 3 (1992): 223.

Bisiach, Edoardo. "Language Without Thought." In *Thought Without Language,* ed. L. Weiskrantz. Oxford: Clarendon Press, 1988, 464-484. The conversation between doctor and patient in which the patient denies ownership of his hand.

Wapner, Wendy, Suzanne Hamby and Howard Gardner. "The Role of the Right Hemisphere in the Apprehension of Complex Linguistic Materials." *Brain and Language* 14 (1981): 15-33. An account of right-hemisphere damaged patients' embellishments of, and alterations to, stories.

Critchley, McDonald. *The Parietal Lobes.* (London: Hafner Press, 1953). A classic book which includes many cases of patients denying ownership of a limb.

Maher, Brendan. "Anomalous Experience and Delusional Thinking: The Logic of Explanations." In *Delusional Beliefs,* eds. Thomas Oltmanns and Brendan Maher, 15-30. New York: John Wiley & Sons, 1988.

Chapter 11

Blackmore, Susan. *Beyond the Body.* London: William Heinemann Ltd., 1982.

Lukianowicz, M. "Autoscopic Phenomena." *A.M.A. Archives of Neurology and Psychiatry* 80 (August 1958): 199-220. The story of the woman seeing her own double upon returning home from her husband's funeral.

Todd, John and Kenneth Dewhurst. "The Double: Its Psycho-pathology and Psycho-physiology." *The Journal of Nervous and*

Mental Disease 122 (1955): 47-55. Guy de Maupassant and his double are featured here.

Ruttledge, Hugh. *Everest 1933.* London: Hodder and Stoughton, 1934.

Chapter 12

Rhodes, Gillian, Susan Brennan and Susan Carey. "Identification and Ratings of Caricatures: Implications for Mental Representations of Faces." *Cognitive Psychology* 19 (1987): 473-497. Evidence for the idea that the brain remembers caricatures of faces.

Sergent, Justine and Jean-Louis Signoret. "Functional and Anatomical Decomposition of Face Processing: Evidence From Prosopagnosia and PET Study of Normal Subjects." *Philosophical Transactions of the Royal Society London B* (1992): 55-62. Images of the progression of face recognition from back to front in the brain.

Sergent, Justine, Brennan MacDonald and Eric Zuck. "Structural and Functional Organization of Knowledge about Faces and Proper Names: A PET Study" *Attention and Performance* XV. New Jersey: Lawrence Erlbaum Associates Inc., 1994. The discovery that biographical information about a name and its face are maintained in separate places in the brain.

Chapter 13

Behrmann, Marlene, Gordon Winocur and Morris Moscovitch. "Dissociation Between Mental Imagery and Object Recognition in a Brain-damaged Patient." *Nature* 359 (15 October 1992): 636-637. The first mention of C.K.

Bornstein, B., H. Sroka and H. Munitz. "Prosopagnosia with Animal Face Agnosia." *Cortex* 5 (1969): 164-169. The man who couldn't recognize his cows.

Diamond, Rhea and Susan Carey. "Why Faces Are and Are Not Special: An Effect of Expertise." *Journal of Experimental Psychology: General* 115 No. 2 (1986): 107-117. Upside-down dogs are hard to recognize too.

Kendrick, K.M. and B.A. Baldwin. "Cells in Temporal Cortex of Conscious Sheep Can Respond Preferentially to the Sight of

Faces." *Science* 236 (24 April 1987): 448-450. The title says it all.

Cohen, Gillian. "Why Is It Difficult To Put Names to Faces?" *British Journal of Psychology* 81 (1990): 287-298. Is it the name or the profession?

Chapter 14

Ekman, Paul. "Facial Expressions of Emotion: An Old Controversy and New Findings." *Philosophical Transactions of the Royal Society London B* (1992): 63-69. Ekman reiterates his theories about emotional faces.

Salk, Lee. "The Role of the Heartbeat in the Relations Between Mother and Infants." *Scientific American* 228 (May 1973): 24-29. Salk's claim that the sound of the heart is the key to holding babies.

Manning, J.T. and A.T. Chamberlain. "Left-Side Cradling and Brain Lateralization." *Ethology and Sociobiology* 12 (1991): 237-244. Manning's claim that it's not the heart, it's the hemisphere.

Grüsser, O-J. "Face Recognition Within the Reach of Neurobiology and Beyond It." *Human Neurobiology* 3 (1984): 183-190. A study of the presentation of the face in paintings.

Chapter 15

Ogden, Jenni and Suzanne Corkin. "Memories of H.M." In *Memory Mechanisms: A Tribute to G.V. Goddard*, eds. Wickliffe Abraham, Michael Corballis and Geoffrey White, 195-215. New Jersey: Lawrence Erlbaum Associates, Inc., 1991.

Scoville, William Beecher and Brenda Milner. "Loss of Recent Memory After Bilateral Hippocampal Lesions." *Journal of Neurology, Neurosurgery and Psychiatry*, 20 (1957): 11-21.

Sherry, David, Lucia Jacobs and Steven Gaulin. "Spatial Memory and Adaptive Specialization of the Hippocampus." *Trends in Neurosciences* 15 No. 8 (1992): 298-303. Chickadees and voles.

Chapter 16

Muter, Paul. "Very Rapid Forgetting." *Memory and Cognition* 8 No. 2 (1980): 174-179. The title says it all.

Baddeley, Alan. "Working Memory." *Science* 255 (31 January

1992): 556-559. A review by one of the leading lights in memory research.

Ellis, N.C. and R.A. Henelly. "A Bilingual Word-length Effect: Implications for Intelligence Testing and the Relative Ease of Mental Calculation in Welsh and English." *British Journal of Psychology* 71 (1980): 43-51. It's harder to remember a string of numbers in Welsh than English.

Daneman, Meredyth and Patricia Carpenter. "Individual Differences in Working Memory and Reading." *Journal of Verbal Learning and Verbal Behavior* 19 (1980): 450-466. Meredyth Daneman's first research paper linking reading comprehension and working memory.

Jonides, John, Edward Smith, Robert Koeppe, Edward Awh, Satoshi Minoshima and Mark Mintun, "Spatial Working Memory in Humans as Revealed by PET." *Nature* 363 (17 June, 1993): 623-625. Watching the visuospatial scratchpad at work.

Ericsson, K. Anders and William Chase. "Exceptional Memory." *American Scientist* 70 (Nov.-Dec. 1982): 607-615. The amazing short-term ("working") memory of S.F.

Bennett, Henry. "Remembering Drink Orders: The Memory Skills of Cocktail Waitresses." *Human Learning* 2 (1983): 157-169.

Chapter 17

Hartry, Arlene, Patricia Keith-Lee and William Morton. "Planaria: Memory Transfer Through Cannibalism Reexamined." *Science* 146 (October 1964): 274-275. The little worms don't really learn better after eating their learned colleagues.

Weaver, Charles A. III. "Do You Need A Flash to Form a Flashbulb Memory?" *Journal of Experimental Psychology: General* 122. No. 1 (1993): 39-46. The brain records even the trivial if asked to.

Brown, Roger and James Kulik. "Flashbulb Memories." *Cognition* 5 (1977): 73-99. The first time the term flashbulb memory was used.

Usher, JoNell Adair and Ulric Neisser. "Childhood Amnesia and the Beginnings of Memory for Four Early Life Events." *Journal of Experimental Psychology: General* 122 No. 2 (1993): 155-165.

Re-establishing the limits of childhood amnesia.

Chapter 18

Luria, Aleksandr. *The Mind of a Mnemonist*. London: Avon, 1968. Shereshevsky's feats.

Treffert, Darold. "The Idiot Savant: A Review of the Syndrome." *American Journal of Psychiatry* 145 No.5 (1988): 653-72.

Loftus, Elizabeth. "The Reality of Repressed Memories." *American Psychologist* 48 No. 5 (1993): 518-537.

Chapters 19 & 20

Winson, Jonathan. *Brain and Psyche*. New York: Random House, 1986.

Hobson, Allan. *The Dreaming Brain*. New York: Penguin Books, 1990.

Antrobus, John. "Bizarreness in Dreams and Waking Fantasy." In *The Neuropsychology of Sleep and Dreaming*, eds. John Antrobus and Mario Bertini. Hillsdale, New Jersey: Lawrence Erlbaum Associates, Inc., 1992.

Smith, Carlyle. "REM Sleep and Learning: Some Recent Findings." In *The Functions of Dreaming*, eds. A. Moffitt, M. Kramer and R. Hoffman. Albany: State University of New York Press, 1993.

Index

Abbott, Edwin, 47
Acetylcholine, 37, 38
Adrianov, Oleg, 14
Alzheimer's disease, 185
American Psychologist, 218
Amygdala, 160
Andreasen, Nancy, 14
Anstis, Stuart, 134
Antrobus, John, 240-241
Asomatagnosia, 104, 107
Astral projection, 118-119
Atropine, 38
Attention, 54-59
Australopithecus africanus, 21

ba, 4
Baddeley, Alan, 181
Battersby, William, 58
Baxter, Doreen, 58
Behrmann, Marlene, 90
Belladonna, 38
Bennett, Henry, 188-189
Berger, Hans, 22-23
Bioscience, 11
Bisiach, Edoardo, 44-45, 46, 48, 49, 53, 62-63
Black, Dr. Sandra, 104, 106, 108
Black-capped chickadee, 169
Blackmore, Susan, 116, 117, 119, 120, 125
Brain stem, 73, 239
Brown, Roger, 199-201
Bryson, Bill, 42
Burning House, 60-62, 63-64, 243

C.K., 139-141
Canadian Brain Tissue Bank, 13
Carey, Susan, 143

Caricatures, 132-133
Carpenter, Patricia, 183, 184
Cathedral, 42
Central executive, 180, 183-185
Cerebral cortex, 10-11, 13, 73, 74, 86, 99, 160, 161, 192
Cerebral hemispheres, 10, 77
Cerebrum, 9, 12
Childhood amnesia, 199, 202-205
Chomsky, Noam, 141
Cobra, 37, venom, 37, 38
Cohen, Gillian, 147
Commissurotomy, 67
Confabulation, 64
Consciousness Explained, 90-91
Convergence zones, 194
Corbetta, Maurizio, 56
Corkin, Dr. Suzanne, 162-167, 171
Corpus callosum, 67
Crick, Francis, 233-234, 236, 238
Critchley, McDonald, 124-125

D.D., 188
D.F., 80, 81
Damasio, Antonio, 193, 194
Daneman, Dr. Meredyth, 178, 183, 184
Dart, Raymond, 21
Dekoninck, Dr. Josef, 220
Delusion, 106-114
Dennett, Daniel, 90
Diamond, Rhea, 143
Dipietro, Vincent, 127
Dopamine, 36-37,
Dreams, 220-231; bizarreness of, 233, 239, 240-241; forgetability of, 233; dreaming to forget, 233-234; dreaming to remember, 234-237; daydreaming, 241, 242

Echidna, 234, 236
Egyptian, 4
Ehrenwald, Jan, 118
Einstein, Albert, 14
Ekman, Paul, 149, 156-157
electroencephalograph (EEG), 22, 23-24, 25, 28, 31, 35, 121, 159, 233, 241
English, 181-182
Enzymes, 34
Epilepsy, 24-25, 26-27, 67, 86, 125, 158-159, 160, 173

face perception, 148-157
Face recognition system, 133-136, 140, 141, 143-144, 145; of infants, 131, 133; birdwatcher, 141; cows, 142; dog experts, 143-144; sheep, 144-145; poodle's head, 146
Falk, Dean, 21
False memory, 217; syndrome, 218
Far-distant space, 76
Flanders, Martha, 79
Flashbulb memories, 199-202;

novelty, 200; consequentiality, 200; bombing of Iraq, 202

Flatland, 47-48

Fore people, 3

Forgetting, 198-199, 207-209, 213, 214, 215

Fraser, Sylvia, 219

Freud, Sigmund, 204, 220, 222, 223, 224, 226, 230, 231, 237, 238, 242; manifest dream content, 222; latent dream content, 222; dream sensor, 222, 223; condensation, 222-223; symbols for sex, 223; day residue, 224

Frogs, 77-79

Frontal lobe, 9, 185, 186-187, 205, 236, 239

Gall, Franz Joseph, 18

Galton, Francis, 214-215

Galvani, Luigi, 22

Garneau, Mark, 91

Gazzaniga, Micheal, 66, 67, 68-69, 70

Goodale, Mel, 79-81

Green, Celia, 227

Grobstein, Paul, 77

Grosskurth, Chris, 209

Grsaping space, 74

Grüsser, Otto-Joachim, 75, 145-146, 152-153

Gyrus, 9, 12

H.M., 158-175, 176-178, 185, 191-192

Haleem, Mohamad, 1,2, 4, 7-8, 14-15

Halligan, Peter, 60, 61, 63

Harvey, Dr. Thomas Harvey, 14

Hippocampus, 136, 137, 160, 161, 168-169, 172, 173, 174, 175, 177, 185, 192, 204, 212, 235, 236, 237, 238

Hoagland, Richard, 128-129

Hobson, Allan, 239-240

Holding babies, women, 153; men, 155

Homo erectus, 21

Hynes, Mrs., 105-106, 107, 108, 109, 110-111, 112

Hypnogogic illusions, 225

Information chunk, 178

Institute of Psychophysical Research, 227-228

Internal mental image (map), 52-54, 58-59

Interpreter, 66, 68-69, 70

K.G., 96-97, 98, 99, 100, 101

ka, 4

Kipling, Rudyard, 37

Kossak, 142

Kulik, James, 199-201

Kuru, 3

Lackner, James, 92, 93

Lashley, Karl, 192-193

Lawrence, D.H., 42

Leech, 39

Left hemisphere, 136, 137, 141, 212, 213
Lenin, Vladimir Ilyich, 14
Libet, Benjamin, 87-89
Liguori, Alphonsus, 116, 117, 121
Limbic system, 101, 160
Lippincott's Magazine, 95-96
Livingston, R.B., 200
Lobotomy, 5-7, 172
Loftus, Elizabeth, 28, 215-217, 218
Long-term memory, 191-205, 207, 211; reawakening of, 215
Lucid dreams, 117, 227-228, 229
Luria, Aleksandr, 208

M.B., 173-174
M.M., 27, 28
Magnetic resonance imaging (MRI), 29
Maher, Brendan, 110, 112
Making a face, 156
Mammalian brain, 160
Manning, Dr. John, 154
Mars Observer spacecraft, 129
Marshall, John, 60, 61, 63
Maupassant, Guy de, 122-123
Maury, Alfred, 228-230; mother of, 230
McConnell, James, 197
Meadow vole, 169
Melzack, Ronald, 96, 100-102
Memory, 160-174
Memory span for digits, 187
Memory molecule, 197
Mental map, 72-73, 83, 90
Merriam's Kangeroo rat, 169
Miller, Julie Ann, 11
Miller, George, 178
Milner, Dr. Brenda, 170-171, 173, 174, 176
Mirror-drawing test, 170-171
Mitchell, Dr. S. Weir, 95
Mitchison, Graeme, 233-234, 236, 238
Moby Dick, 94
Modules, 141, 145
Moffitt, Allan, 222
Molenaar, Gregory, 127
Mongoose, 37, 38
Moniz, Egas, 5
Monkeys, 144, 145
Montreal Neurological Institute, 135, 141, 170
Moon illusion, 75-76
Moscovitch, Morris, 90
Moscow Brain Institute, 14
Mutter, Dr. Paul, 179

Narcolepsy, 225, 227
NASA, 129
Neglect, 45-46, 47, 48-50, 51-52, 53, 54, 55, 56, 57-59, 60, 62, 64, 66, 69, 70, 71, 89-90, 104, 107
Neisser, Ulric, 202, 203, 204, 205
Neural nets, 233

Neuron, 31-32, 34, 35, 40, 41, 84, 99, 207, 235
Neurosignature, 101
Neurotransmitter, 32-34, 35, 37, 38-39, 40-41,195, 196, 225, 233, 239
Neville, Helen, 24
New Guinea, 149
New Scientist, 11
Nisbet, Richard, 64-66
Noe, Dr. Adrianne, 15
"Now Print!", 200-201

Occipital lobe, 9
Ogden, Jenny, 167
Ojemann, Dr. George, 12
Out-of-body experience (OBC), 115-126, 227
Owl monkeys, 86

Parietal lobe, 9, 45, 49, 52, 101
Parkinson's disease, 36-37, 98
Pavlov's dogs, 195, 196, 207
Penfield, Dr. Wilder, 24-28, 86, 87, 215
Personal space, 74-75
PGO (pontogeniculooccipital) spikes, 239
Phantom pain, 97-98
Phantom limbs, 95, 96-103
Phobia, 70
Phonological loop, 180, 181, 182, 186, 189
Phrenology, 16-19, 22,
Piazza dei Cinquecento, 49
Piazza del Duomo, 42-45, 47, 52
Planaria, 197
Poggio, Tomaso, 81
Pope Clement XIV, 116, 121
Portrait paintings, 153, 154
Positron emission tomography (PET), 28-30, 31, 56, 57, 186
Preformationism, 83
Prosopagnosia, 130, 135
Prozac, 39-40

R.V., 80, 81
Rapid-eye-movement sleep (REM), 221, 224-225, 226-227, 229, 233, 234, 235-236, 237, 241-242; REM windows, 238
Reading, 183-184
Repressed memories, 217-218
Reverse learning, 233
Right hemisphere, 112-113, 136, 150-152, 154, 155, 212, 213
Rimland, Dr. Bernard, 210-211, 213

S.F., 187-188
Sacks, Oliver, 47, 210
Salk, Lee, 153
Salk Institute, 24
Saturday Night, 219
Savants, 209-213; calendar calculators, 211, 212-213
Scarella, Jacques, 209
Schacter, Daniel, 109

Schilder, Paul, 54-55
Schizophrenia, 35-36, 37, 125, 172
Scientific American, 11, 23
Scoville, Dr. William Beecher, 159, 160, 163, 172-174
Seigel, Ronald, 125
Sensory homunculus, 83-87, 89, 90, 99, 100, 101
Sergent, Dr. Justine, 135, 136, 137, 141
Serotonin, 39-40
Sheep, 144-145
Shereshevsky, 208-209, 211, 213
Sherry, Dr. David S., 169
Short-term memory, 176-180; length of retention of, 179; students and waitresses for drink orders, 188-189
Skull, 16-17, 19-21, 23
Sleep, phase of, 221; lab, 220; paralysis, 225
Smith, Carlyle, 237-238
Smythe, F.S., 124
Soechting, John, 79
Spiny anteater, 234
Split brains, 67-69, 70
Spurzheim, Johan Kaspar, 18
Square, A., 48
Sulcus, 9, 12
Synapse, 32, 35, 36, 40, 195, 196

Tart, Charles, 120
Taung child, 21
Temporal lobe, 9, 27, 52, 130, 134, 136, 159, 160, 172
The Face, 127
The Atlantic Monthly, 94-95, 96
The Journal of Neurosurgery, 160
Theosophical Society, 118-119
Theta waves, 235
Treffert, Dr. Darold, 212

Viking spacecraft, 127, 134
Visual Agnosias, 116-117
Visual imagery, 122
Visual agnosia, 139
Visuospatial scratchpad, 180, 182, 186, 189

Walter, William Grey, 23, 24
Warrington, Elizabeth, 58
Weaver, Charles, 201-202; bombing of Iraq, 202
Weiland, I. Hyman, 153
Welsh, 181-182
Wiff'n'Proof, 237, 238
Wilson, Timothy De Camp, 64-66
Winson, Jonathan, 234-237, 238, 240
Working Memory, 180-187, 190, 191
Worsley, Allan, 228

Yakovlev, Dr. Paul Ivan, 2, 3, 7-8, 15
Yakovlev Collection, 1-3, 5, 7-8, 12, 13, 15
Yakovlev-Haleem Collection, 15